Arthropods - Are They Beneficial for Mankind?

Edited by Ramón Eduardo Rebolledo Ranz

Published in London, United Kingdom

IntechOpen

Supporting open minds since 2005

Arthropods - Are They Beneficial for Mankind?
http://dx.doi.org/10.5772/intechopen.77940
Edited by Ramón Eduardo Rebolledo Ranz

Contributors
Jeyachandran Sivakamavalli, Kiyun Park, Inh-Sil Kwak, Vaseeharan Baskaralingam, Jobi J. Malamel, Gregorius Nugroho Susanto, Sajjalavarahalli Gangireddy Eswara Reddy, María F. Fernanda Barberena-Arias, Elvira Cuevas, Jinu Medhi, Mohan Chandra Kalita, Jintu Dutta, Gyanpriya Maharaj, Godfrey Bourne, Abdullah Adil Ansari, Hendrik Sithole, Nolubabalo Tantsi, Victoria Wojcik, Cordelia Ebenebe, Valentine Okpoko, Bernadethe Ezenyilimba, Maduabuci Amobi, Oghalo Okore, Simon Okweche, Michael Okonkwo, Joan Nneamaka Eze, Geoffrey Maxwell Malinga, Robert Opoke, Karlmax Rutaro

First published in London, United Kingdom, 2021 by IntechOpen
IntechOpen is the global imprint of INTECHOPEN LIMITED, registered in England and Wales, registration number: 11086078, 5 Princes Gate Court, London, SW7 2QJ, United Kingdom
Printed in Croatia

British Library Cataloguing-in-Publication Data
A catalogue record for this book is available from the British Library

Additional hard and PDF copies can be obtained from orders@intechopen.com

Arthropods - Are They Beneficial for Mankind?
Edited by Ramón Eduardo Rebolledo Ranz
p. cm.
Print ISBN 978-1-78984-165-7
Online ISBN 978-1-78984-166-4
eBook (PDF) ISBN 978-1-83880-752-8

For EU product safety concerns:
IN TECH d.o.o., Prolaz Marije Krucifikse Kozulić 3, 51000 Rijeka, Croatia, info@intechopen.com or visit our website at intechopen.com.

Meet the editor

Ramón Eduardo Rebolledo Ranz received an Agricultural Engineering degree from the Austral University of Chile in 1986 and a Doctor of Agricultural Engineering with a mention in Plant Protection from the Polytechnic University of Madrid in 1994. He has worked on more than seventeen research projects on agricultural entomology, biodiversity, and beekeeping. He has published eighty-five scientific articles in national and foreign specialty journals. He has written one book and five book chapters in his specialty. In addition, he has been editor of four books on applied entomology. He has presented more than 100 works in different national and international scientific congresses on entomology and beekeeping. He has directed more than eighty undergraduate and graduate degree theses. He is a member of the scientific communities of beekeeping and entomology and has continued to organize more than twenty scientific congresses and seminars in his specialty. He is a reviewer for scientific journals and books. He has been president and director of different scientific societies and advisor to the Chilean Beekeeping Network, where he is also a consultant to the Latin American Beekeeping Federation for the congresses held in different countries. He is also an advisor to private companies in the agricultural sector on beekeeping and pest control issues.

Contents

Preface

Arthropods are the most abundant and diverse group of invertebrate animals in existence. They have many purposes including pollination, food for both humans and domestic animals, bioindication, biological pest control, and more. Unfortunately, like all species of animals and plants existing in the world, arthropods face serious conservation problems and the risk of extinction.

This book covers topics such as arthropods as food for humans, arthropods as a bio-indicator species, use of arthropods in the cosmetic and pharmaceutical industries, arthropods and their conservation status, diets for raising arthropods, models that relate arthropods to climate change, enemies that biologically control some insect pests, and the coloration of flowers and its relationship with pollination by butterfly species. It is a useful reference for undergraduate and graduate students, academics, researchers, and anyone interested in learning more about this important group of invertebrates and their current conservation status.

It has been a great pleasure serving as editor of this volume. I thank all contributing authors for their effort and participation.

Ramón Eduardo Rebolledo Ranz
Universidad de La Frontera Casilla,
Temuco, Chile

Chapter 1

Arthropods in Cosmetics, Pharmaceuticals and Medicine: A Review

Cordelia Ebenebe, Simon Okweche, Oghale Okore, Valentine Okpoko, Maduabuchi Amobi, Joan Nneamaka Eze, Benedeth Ezenyilimba and Michael Okonkwo

Abstract

Apart from food, other important needs in the care of human bodies are cosmetics and drugs. For long the latter two are obtained from chemical formulations and phytochemicals (commonly used in Ethnomedicine), use of bioactive compounds from insects (i.e. "ento medicine" and "ento cosmetics") is a recent development in research, even though the bioactive compounds were discovered long ago. This chapter is a review on a number of substances extracted from various insect species that are useful in cosmetics, pharmaceutical industries as well as those that form part of prescription for healing in orthodox and traditional medicine. The review is based on information from scientific reports, Google, e-library, textbooks. A number of substances were found to have been incorporated into cosmetic and pharmaceutical products and as part of prescriptions for healing in orthodox medicine, many others at elementary stages of investigation, purification and development. The findings showed that insects have a lot of bioactive substances that need to be harnessed for the good man.

Keywords: arthropods, uses, cosmetics, pharmaceuticals, medicine

1. Introduction

Arthropods, a word coined from two Greek words: "arthron" meaning "joint" and "pous"(podos) meaning "foot" (the two together means "jointed legs) comprise all animals with jointed legs including insects, spiders, mites, ticks, and scorpions. Radis-Baptista and Konno [1] stated that the myriad of animals under this phylum are grouped into four subphyla namely Chelicerata, (Arachnids), Crustacea, Myriapoda (Centipedes) and Hexapoda (insects). A lot of these arthropods produce some chemical substances and/or venom to protect themselves from danger of being killed or being captured, while some produce substances for yet to be identified purposes. Radis-Baptista and Konno [1] noted that thousands of arthropods species ranging from arachnids (spiders and scorpions), to hymenopterans (ants, bees and wasps) and myriapods (centipedes) are venomous and utilize their venoms for chemical ecological warfare that includes individual and colonial defense, predation and paralysis of coexistent species to nourish their brood. The arthropods inject their self-produced venom or other substances into their victims through bites, stings or spray of aerosol-like chemicals. Scientists have shown [2–4] that these arthropods secretions/venoms may include toxins (e.g., inorganic or organic compounds, alkaloids, etc.) and other compounds (e.g., histamines, enzymes) [5], that may enhance the effectiveness or "spread" of the toxins. These arthropod secretions (whether venom or other substances) contain a number of chemical substances useful for cosmetics, pharmaceutical and nutraceutical products, as well as medicine. This paper is a review of chemical substances and venom of some arthropods useful in the cosmetic, pharmaceutical industries as well as those utilized in orthodox and traditional medicine.

2. Arthropods and cosmetics

2.1 Cochineal or carmine (Hemiptera: true bugs)

Cochineal or carmine is a red insect dye from female scale insect *Dactylopius coccus*, a cactus-eating insect native to tropical and subtropical regions of South America and Mexico. It was earlier used for dying cloth by the Aztecs, but today, it is used as a dye in foods, cosmetics, drugs, and dyeing of textiles. The red pigment (carmine) is made from crushing female cochineal beetles, and it reportedly takes 70,000 to 100,000 dead females to produce 1kg of cochineal dye [Business insider 2011]. The female cochineal beetles gain a red colour through their diet of red berries, and so produce red carminic acid when threatened. In the cosmetics industry, cochineal is used to dye lipstick, blush, and eye shadow. It creates bright, bold and deep red colors. Today, the cochineals are harvested mainly in Peru and the Canary Islands on plantations of prickly pear cacti which is the preferred host of the bug. The natives sun-dry the insects, crush it and dunk in an acidic alcohol solution to produce carminic acid, the pigment that eventually becomes carmine or cochineal extract, depending on the processing method adopted. A number of works [6], have been going on how to increase the yield of carmine from the *Dactylopius coccus.*

2.2 Lac and shellac from scale insects (Hemiptera)

Lac is a the resinous secretions from a number of species of lac scale insects, *Laccifer lacca,* and most of it is produced in India, Lac is an important ingredient of many items, including floor polishes, shoe polishes, insulators, various sealants, printing inks, and varnish. In the cosmetic industry, it is used mainly for nail polish remover. Juliane [7] described the three processes involved in the production of shellac from lac. According to him, cultivation begins when the farmer gets a stick containing eggs ready to hatch and ties it to the tree to be infested. Thousands of lac insects colonize the branches of the host tree and secrete the resinous substance, the coated branches are cut out and harvested as sticklac. The harvested sticklac is sieved and subjected to a number of washings to remove impurities of all kinds, resulting in a purer form called seedlac. However, seedlac still contains 3-5% of impurity, further removal of impurities by heat treatment or solvent extraction results in the product called shellac. Shellac is therefore a processed secretion of the lac insect.

Raman [8] described shellac as a resin secreted by the female lac bug on trees in the forest of India and Thailand, processed and sold as flakes and dissolved in alcohol to form liquid shellac. *Tachardia lacca* or *Kerria lacca.* In the cosmetic industry, the shellac is used for personal care products in hair sprays, eyeliners and mascara. Das [9] stated that shellac is a common ingredient in cosmetics such as mascara, lipsticks, nail polish and hairsprays. This is because shellac has the ability to hold the ingredients together in a compressed tablet or cake. It also keeps an emulsion from getting separated into water-soluble and oil soluble components, so it helps the hair hold its style by inhibiting the hair's ability to absorb moisture (i.e. shellac is non hygroscopic in nature). The Cosmetic Ingredient Review (CIR) panel has assessed shellac's safety and concluded that cosmetic-grade-shellac is safe for use in personal care and cosmetic product formulations up to 6%. From 1997 to 2002 Ken Golz-Berner has patented seven cosmetic products containing Shellac in various percentages.

2.3 Insect oil in cosmetic uses

2.3.1 Cosmetic oils from some dipterans

Rebecca Guenard a columnists writing in Olio, an Inform column of Inform Magazine that highlights research, issues and technologies of interest in fats and oil stated that fats and oil are major components of cosmetics, she further stated that linolenic acid heals dryness while triglycerides soften the skin. According to her report, skin care cream formulators earlier depended on mink oil as a source of these ingredients, but ethical considerations led to a shift to Macadamia nut oil with a similar fatty acid profile. However, competition with food use also limited Macadamia oil application in cosmetics thus the current interest in insect oil. A team of researchers from Thomas Moore University College and University of Antwerp both in Belgium assessed the usefulness of oil from three insect species: Black soldier fly (BSF) (*Hermetia illucens*), locust (*Locusta migratoria*) and the house cricket (*Acheta domesticus*) as an alternative source for the production of fats for cosmetics. The research was reported in [10] and they discovered that insect fats contained shelf life limiting phospholipids, which they were able to remove with a degumming procedure. After a gamut of tests and trials, the fats after extraction and refining were used in hand cream formulations and compared with conventionally used mink and plant derived oils. The fatty acid analysis showed that BSF contains >60% of lauric acids making it less suitable as a skin care product, whereas locust and cricket fats are rich in C16 and C18 fatty acids which makes them more suitable. However, the phospholipids and free fatty acids were found to be higher compared to commercially refined oils so they need to be removed by appropriate refining protocols, odor and colour also need to be removed for better applicability. They therefore concluded that with further refining locust and cricket oil would lose their colour and odour and become useful as a cosmetic product. Recently, a US company launched the world\s first youth generating face oil containing insect oil extract by name Point68. The oil was developed as a joint venture between SIBU ® and insect industry professional Josh Galt. Point 68 is reported to be a luxury face oil formulated to improve skin hydration with cellular healing and rejuvenation.

2.3.2 Honey and beeswax as cosmetic resources (Hymopterans)

Ediriweera and Premarathna [11] gave a long list of cosmetic uses of honey and beeswax in the beauty industry as a skin moisturizer, softener and a healer of skin tissue.

- **Face wash**: A small quantity of lemon juice is mixed with 5ml of bee's honey and applied to the face before washing. This is used as a home remedy

- **Facial cleansing scrub**: 5g of almond seed powder is mixed with 5ml of bee's honey, scrub softly and then wash [12]

- **Facial smoothness improver**: A tablespoon of honey is whisked together with white of an egg, 1 teaspoon of glycerine and 1/4 cup of flour makes an excellent firming mask. Just smooth on the face, leave on 15 min, and rinse off with warm water [12]

- **Facial softness improver**: One or two tablespoons of honey is mixed with one-third cup finely ground oatmeal. A teaspoonful of rose water is then added. This is used to clean the face thoroughly. This is spread evenly over the face. Allow to stay for 10 min to 1.5 hrs. Clean off with a soft washcloth and warm water, then rinse with cold water

- **Facial moisturizing pack**: 2 tablespoons of honey is mixed with 2 teaspoons of whole milk. This is applied over the face and allows it to stay for 15mins. It is then rinsed off with warm water, and then with cold water.

- **Pimples:** Honey is applied direct on pimples

- **Cracked lips**: Honey is applied on cracked lips

- **Lotion for dry patches of skin**: 5ml of bee's honey is mixed with 5ml of olive oil and 2.5 ml of lemon juice. It is applied on the skin and washes after 15 min.

- **Hair lustre**: 5ml of bee's honey is mixed with 4 cups of warm water. It is used as a hair rinse.

- **Conditioner:** 10 ml of olive oil is mixed with 5ml of bee's honey and applied on hair. It is washed after 15 min.

Burlando and Cornara [13] opined that in cosmetics preparations, honey exerts emollient, humectant, soothing and hair conditioning effects, while keeping the skin juvenile, retarding wrinkle formation, regulating pH and preventing pathogenic infections. According to them, honey based cosmetics products include lip ointment, cleansing milks, hydrating creams, after sun, tonic lotions, shampoos and conditioners ranging from 1 to 10% honey inclusion, though some products can contain up to 70% honey when mixed with oils, gels and emulsifiers or polymer entrapments.

2.3.3 Sericin from silkworm (Lepidoptera): a natural moisturizer in cosmetic products

Silkworm moth (*Bombyx mori*) is a lepidopteran, whose caterpillar is involved in production of silk during the process of cocoon formation. Silk is made up two proteins: fibroin (70-80%) and sericin (20-30%). Fibroin is the structural and fibrous part of the silk, while sericin is the gummy coating that gums the fibres and makes them stick to each other. Kunz *et al.* [14] discussed extensively on the use of sericin in cosmetics formulations such as creams and shampoos. Voegeli *et al.* [15] reported that inclusion of sericin in creams lead to hydration, elasticity of the skin thus leading to anti- aging and anti-wrinkle effects. Yamada *et al.* [16] also reported sericin in creams help to prevent nails from chapping and brittleness.

These applications are especially due to presence of amino acids with hydrophilic side groups (80%) such as serine (30-33%) which has large capacity to absorb water. Padawar *et al.* [17] cited in [14] studied in-vivo moisturizing effect of sericin on human skin and found its action to decrease the impedance and increase in the level of hydroxyproline and hydration of epidermal cells. The increase in hydration was attributed to the occlusive effect of sericin which prevented the transepidermal water loss, responsible for skin dryness. Sharma [16] stated that industrial production of one ton of silk yields 200kg of sericin and that sericin is isolated from cocoons of three types of silkworm: *Bombyx mori* (Mori), *Antheraea assamensis* (Muga) and *Philosamia ricin* (Eri). According to her Muga sericin embedded cream could be used for topical skin care application as a potential therapeutic to protect skin against UV radiation –induced inflammation, oxidative damage of epidermal keratinocytes, aging, wrinkling, preventing skin roughness, enhancing the skin elasticity and moisture [18, 19].

2.4 Arthropods and pharmaceuticals

2.4.1 Antimicrobial peptides (AMP) in wasp, ants and other arthropods

Insects are major sources of antimicrobial peptides/proteins (AMPs) which are cationic and comprise less than 100 amino acids [20–22] (**Table 1**). Their AMPs enable them to form resistance against bacterial infections. Antimicrobial peptides are short immunity related proteins that can fight bacteria, viruses, fungi and parasites [25] (**Figure 1**). According to them, AMP are secreted from cells and tissues that contribute to host innate immunity such as the haemocytes or the fat body and could be a valuable alternative antibiotics especially in this era of growing antimicrobial resistance [22, 25].

Dipteran species	**Maggot factors**	**Activity against**
Sarcophaga peregrina	Sarcotoxin 1A	Gram-negative bacteria
Sarcophaga peregrina	Sapecin B , Defensin A	Gram-positive bactéria, MRSA, fungi, cancer cells
Drosophila melanogaster	Cecropin A, B, C	Gram-negative bacteria, Gram-positive bactéria, fungi
Phormia terranovae	Defensin A	Gram-positive bacteria
Lucilia sericata	Lucifensin	Gram-positive bacteria, MRSA
Calliphora vicina	Alloferon 1	Viruses and anti-cancer with cytokine activity
Lucilia sericata	p-hydroxybenzoic acid p-hydroxyphenylacetic acid Octahydro-dipyrrolo [1,2-a;10,20-d] [1,2-a;10,20-d]	Bacteria
Lucilia sericata	Seraticin	Bacteria, MRSA
Sarcophaga peregrina	5-S-GADa	Bacteria and anti-cancer

Source: Yi et al, [23, 24].

Table 1.
Anti-bacterial and anti-cancer factors produced by Dipteran larva.

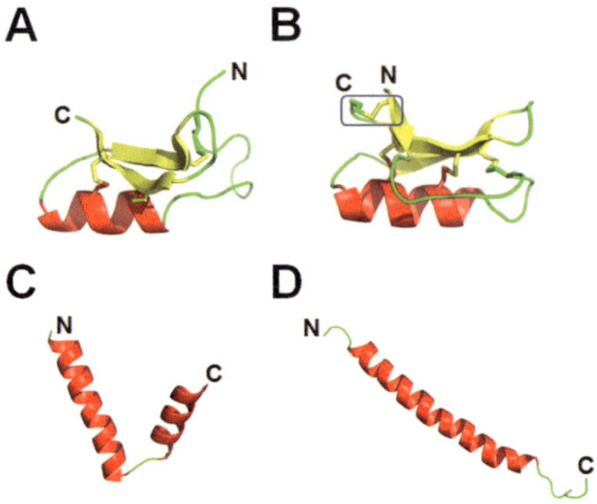

Figure 1.
Structures of an insect defensin, drosomycin, cecropin, and moricin. (A) Protophormia terraenovae defensin *(B) Drosophila* melanogaster drosomycin *(C) Papilio xuthus papiliocin (cecropin) (D) Bombyx mori moricin. Source: Yi et al, [23].*

Reported on a number of antimicrobial substances found in insects [26]. The examples they gave included alloferon, an antimicrobial compound produced by blow fly larva, used as an antiviral and antitumor agent in South Korea and Russia and a compound sourced from the venom of the wasp *Polybia paulista* capable of killing cancer cells without harming normal cells. Ant venom can inhibit bacteria such as *Enterococcus faecalis* and *Listeria monocytogenes.* The types of antimicrobial peptides isolated from ants vary between ant species (e.g., polusulin I and II from the Jack jumper ant and solenopsins from the fire ant). In ancient times and today in some parts of Mexico, venom from red harvester ants is used to treat arthritis-like diseases (i.e., ants were placed on the affected areas and allowed to sting) [27].

Attacins another antimicrobial substance found in reasonable amount in *Heliothis virescens, B. mori* and *Hyphantria cune* that are active against gram negative bacteria such Gram-negative E. coli and *Citrobacter freundii* as well as the fungus *Candida albicans* [28]. Moricin was first isolated from *B.mori* larva, and can be found in other lepidoptera insects. Moricins have activity against Gram-negative and Gram positive bacteria [29], *Galleria mellonella* also show high activity against filamentous fungi and yeast [30].

The antimicrobial substance in wasp venom is mastoparan and it is found to be highly effective in decimating microbes, however, the high degree of damage to mammalian cells has prevented its use as a therapeutic compound. Other compounds found in wasp venom (e.g., eumenine-mastoparan, aumenitin, protonectin, paulistine) are being explored for their antibacterial properties and seem to have a minimal effect on mammalian cells.

Apart from insects, antimicrobial substances have also been extracted from the venom of other arthropods. A good example is lysosin-I from wolf spider (*Lycosa singoriensis*) venom which can inhibit bacteria such as *Shigella dysenteria* and *Staphylococcus aureus.* Budnik *et al* [31] stated that *Lycosa singoriensis*contain antimicrobial properties (lycocitin 1, 2, 3) that can inhibit the growth of gram positive and gram negative bacteria. Another important one is scolopin I, scolopin II, scolopendrin I (antibacterial peptide compounds) found in Chinese red-headed centipede [2].

Other insects with antimicrobial properties and the major molecules listed in [32] including:

Insect Molecule with AMB	Property	Author
Ixodes persulcatis	Persulcatusin	[33]
Spodoptera Kitura	Cecropin like peptides	[34]
Tetramorium bicarinatum	Bicarinalin	[35]
Callifora vicina	Alloferon	[36]
Hermetia illucens	Trx-Stomoxyn ZH I	[37]
Galleria mellinella	Apolipophorin III	[38]

2.4.2 Insects with angiotensin converting enzyme (ACE) inhibitors

Angiotensin Converting Enzyme (ACE) is a widely distributed zinc metallopeptidase that represents the final enzymatic step in the lysis of angiotensin I to produce angiotensin II. Enzyme (ACE) is a central component of the renin-angiotensin system (RAS) which controls blood pressure by regulating the volume of fluids in the body i.e. it converts the hormone angiotensin I to the active vasoconstrictor angiotensin II. According to Riordan [5] ACE I is a monomeric, membrane bound zinc and chloride dependent peptidyl dipeptidase that catalyzes the conversion of decapeptide angiotensin I to the octapeptide angiotensin II by removing a carboxyl terminal dipeptide. Riordan [5] also stated that ACE also known as peptidyl-dipeptidase A or Kininase II was first isolated in 1956 and shown to be to be chloride dependent metalloenzyme that cleaves dipeptide from the carboxyl terminus of the decapeptide angiotensin I to form a potent vasopressor (blood vessel constrictor) angiotensin II.

Herman et al. [39] reported the 2014 evidence based guideline by the Eight Joint National Commission recommended ACE inhibitors as one of the four drugs for the treatment of high blood pressure. Cito *et al.* [40] listed such synthetic ACE inhibitor drugs used to include captopril, lisinopril and ramipril, for the treatment of cardiovascular diseases and hypertension, but their use is often associated with some side effects, such as hypotension, coughing, loss of taste, reduced renal function and angioedema [41]. Therefore, natural compounds such as ACE inhibitor peptides derived from food sources such as insects with little or no side effects may replace or support antihypertensive drugs. According to Cito *et al* [40] ACE inhibitory activity has been detected in protein hydrolysates from insect species belonging to the orders Coleoptera, Diptera, Hymenoptera, Lepidoptera and also Orthoptera. Further investigations by the team led to identification of specific ACE inhibitory peptide from silkworm *Bombyx mori* (Lepidoptera: Bombycidae), the yellow mealworm Tenebrio molitor (Coleoptera: Tenebrionidae), the cotton leafworm Spodoptera littoralis (Lepidoptera: Noctuidae) and also from the weaver ant Oecophylla smaragdina (Hymeoptera: Formicidae).

2.4.3 Fibroin and sericin from silkworm and their therapeutic potentials

Silkworm (*Bombyx mori*) larvae have two proteins necessary for cocoon formation: fibroin and sericin. Fibroin forms the fibrous weave of the cocoon and sericin a coextruded coating that acts as a matrix in the resulting non-woven composite cocoon [42]. Kunz [14] showed that the presence of hydrophobic amino acids and its antioxidant potential make it possible for sericin to be applied in the food industry and cosmetic industry. Furthermore, bioactive substances found in fibroin and sericin, exhibited a therapeutic potential for the reduction of plasma glucose level. Ryu *et al.*, [43] reported that silkworm powder has blood glucose lowering effects induced by uptake of freeze-dried 3rd day of 5th instar silkworm larval powder indicating that consumption of silk worm may assist in controlling glucose level.

Seo [44] studied the antihyperlipidemic and body fat lowering effects of silkworm proteins with different fibroin/sericin compositions in mice fed with a high fat diet. They noted that the hypolipidemic effect was partly due to increased fecal lipid excretion, inhibition of lipogenesis and regulation of adipokine production. Their findings illustrate that silk protein particularly sericin may be beneficial in the prevention of high fat diet-induced hyperlipidemia and obesity. Seo *et al.* [44] also reported the anti-adipogenic and antiobesity effects of *T. molitor* larvae in vitro and in vivo. The results showed that the daily intake of yellow mealworm larvae powder by obsessed mice attenuated body weight gain by reducing lipid accumulation and triglyceride content in adipocytes, thus indicating the potential of insect bioactive compounds to induce weight loss.

Martinez-Mora *et al.* [45] stated that fibroin and sericin in silkworm *Bombyx mori* are used to stimulate wound healing, as they bring about the generation of a basal epithelium capable of replacing the epidermis of the wound. Aramwit and Sangcakul [46] stated that sericin is a wound healing agent. Ersel *et al.* [47] also reported the positive influence of sericin in the healing of incision wounds and therefore recommended sericin – based formulations in the treatment of incision wounds.

2.4.4 Antioxidants

Antioxidants also called free radicals scavengers are substances that can prevent or slow damage to cells caused by free radicals (unstable molecules that the body produces in response to environmental and other pressures) that has damaging effects on body cells. The sources of antioxidants can be natural or artificial and both are potentially beneficial in the prevention of cardiovascular and other diseases associated with free radicals in the body [48]. Felton and Summers [49] posited that insects possess a suite of antioxidant enzyme and small molecular weight antioxidants that may form a concatenated response to an onslaught of dietary and endogenously produced oxidants. The antioxidant enzymes of insects are generally composed of Catalase (CAT) and Peroxiredoxin (Prx) [50]. Antioxidant enzymes such as superoxide dismutase, catalase, glutathione transferase, and glutathione reductase have been found in insects while the roles of ascorbate, glutathione, tocopherols, and carotenoids has not being well studied in insects but may play very important antioxidant roles [49]. Roos and Van Huis [48] argued the postulations that antioxidants in food have a positive, because they opined that the activity of antioxidants is likely to change during digestive and metabolic processes. EFSA [51] stated that antioxidants in food including insects need to be tested in vivo in humans before claiming its benefits.

Antioxidants are also used as additives that preserve food from "farm to plate" and militate against oxidative deterioration of foods during processing and storage. Due to their high stability and low volatility, the antioxidants help to maintain the level of nutrients, the texture, color, taste, freshness, functionality, aroma, and appeal to consumers such as the older person, ceteris paribus. The possibility of extracting and using antioxidants found in insects for this purpose is yet to be reported in literature.

2.4.5 Chitin and chitosan

Chitin according to [52] is a macromolecular compound, biopolymers found in insects and crustaceans with high nutritional and health benefits. Islam *et al.* [53] described chitin as a natural mucobiopolymer, hard, white, inelastic and nitrogenous compound, composed of randomly distributed N-acetyl-D-glucosamine (N-G1cNAc) monomers. Chitin was first isolated from the cell walls of mushroom by French chemist Henri Braconnot in 1811, and he named it "fungine" before Odier (1823) renamed it chitin. Tharsnathan and Kittur [54] noted that chitin is the second most abundant biopolymer on earth. Within annual biosynthesis of more than 100 billion tons. The exoskeleton covering the body of insects, crustacean ns and some annelids contains polysaccharide chitin which accounts for 5–20% of their dry weight [55, 56]. Chitin is the second biggest available biopolymer on earth, next to cellulose and it is a primary component of the exoskeletons of arthropods [48]. Chitin serves as a raw material for making suture and wound dressing. Industrially, chitin and its products are used as a raw material for the production of hair care and skin products [57, 58]. Chitosan, though found in some types of fungi but most are obtained from chitin deacetylation. Chitosan shows more versatility than chitin due to its solubility and reactive free amino group. Chitosan is used in the food industry because they are non-toxic for warm blooded animals. It is also used as an emulsifying and gelling agent, to stabilize food. Microcrystalline chitin, addresses the problem of toxic substances consumption as it is used as flavor, colourants and dietary fibre in baked food [59].

2.4.6 Apitoxin (bee venom) and melittin

Apitoxin or honey bee venom according to [60] is a cytotoxic and hemotoxic bitter, colorless liquid containing proteins which may cause local inflammation (**Table 2**). Lee and Bae [4] noted that bee venom constituents include amphipathic polycationic peptides of which major ones are melittin and apamin, enzymes such as phospholipase A2, and low molecular weight compounds including active bioamines such as histamine and catecholamines, while Hoffman [61] opined that

Insect	Component	Activity
Honey bee, wasp	Melittin peptide	Kills bacteria and cancer cells, anti-inflammatory
Bumble bee	Bombolitins peptide	Antimicrobial
Honey bee, wasp	Apamin peptide	Treat muscular dystrophy and kill tumours
Honey bee, bumble bee	Mast cell degranulation peptide (MCD)	Analogues inhibit allergies
Honey bee, bumble bee, wasp	Hyaluronidase enzyme	Enhance cancer chemotherapy
Honey bee	Adolapin and other polypeptides	Analgesic and anti-inflammatory
Honey bee bumble bee, wasp	Phospholipases A1	Kills cancer cells and inhibits malaria
Wasp	Phospholipases A1	Sting diagnosis and immunotherapy
Wasp, bee	Kinins e.g. bradykinin and other neurotoxins	Pain control and neurological diseases
Beeswax, honey & royal jelly	Bio-active, oestrogenic and immunosuppressive compounds	
Bee venom	Anti-malignant, anti-inflammatory, anti-arthritic peptides, e.g. melittin	Kill cancer cells
Serrapeptase: "The miracle enzyme"		

Table 2.
Bioactive compounds in bees and wasps.

apitoxin (bee venom) is a complex mixture of several biologically active proteins and neurotransmitters such as phospholipases A2 and B, hyaluronidase, serotonin, histamine, dopamine, noradrenaline, and adrenaline, some of which can contribute to the clinical signs and symptoms of envenomation. Many authors have reported effective use of apitoxin in the treatment of a number of diseases including Oršolić [62], skin conditions [63] and even Parkinson's disease [64]. Besides, purified bee venom preparation called ApitoxR has been approved by the FDA as a subcutaneously injectable product for relieving pain and swelling associated with rheumatoid arthritis, tendinitis, bursitis, and multiple sclerosis [65].

Melittin, the major component of apitoxin about 40-50% of its total dry weight [66] is a 26 amino acid long amphipathic peptide. The 26 amino acids in melittin have an aminoterminal region that is hydrophobic, and a carboxylterminal region that is hydrophilic. This feature according to [67] enables melittin to accumulate in cell membranes, disrupting their phospholipid backbones and causing cell lysis. Melittin not only induces the lysis of a wide range of plasmatic membranes but also of intracellular. Oršolić [62] is of the view that the antibacterial effects and potential anticancer efficacy of this membrane-active peptide may be related to cytolysis following the activation of phospholipase A2, the induction of pores or perturbations in plasma and subcellular membranes, and the activation of apoptotic pathways such as those mediated by caspases and metalloproteinases membranes such as those of mitochondria.

The silkworm (*Bombyx mori*) is reported to harbor an enterobacterium of Serratia species which produces a substance called serrapeptase with tremendous pain relieving potential and therefore useful in reducing inflammation and pain without the usual side effect of synthetic drugs. According to [68] posited that serrapeptase also called serratiopeptidase is a proteolytic enzyme with many biological properties like anti-inflammatory, analgesic, anti-bacterial, fibrinolytic properties, thus its wide use in clinical practice for treatment of many diseases. The silkworm miracle enzyme is actually produced by a friendly bacterium found within the silkworm intestine known as Serratia E15. Serratia E15 protease proteolytic properties is responsible for dissolving the cocoon for the emergence of the moth i.e. it dissolves the casing (cocoon) which houses the pupa, for the pupa to metamorphose to adult and therefore facilitate the emergence of the adult. Proteolytic enzymes are naturally produced by the human body and other living organisms [69] and include peptidases, proteases and proteinases which are essential for many important processes in the body. They govern all of the body's metabolic functions and regulate the functioning of the body's proteins. They also play an important role in protein digestion, immune function and other vital functions.

According to the article "Silky Solution to Heart Disease" written by Wellness watchersMD, many people in recent times are deficient of proteolytic enzymes because of poor diet and unhealthy eating habits and so the body is unable to properly digest and assimilate processed foods. The result is that, as the body uses up its reserve of these enzymes, it diminishes the amount of the enzymes that are available to oversee many metabolic functions for which proteolytic enzymes are responsible; resulting in unhealthy consequences such as

- The development of chronic, low-grade inflammation, which scientists and physicians now know is one of the primary causes of all degenerative disease.
- The accumulation of waste and debris in the blood and lymphatic system.
- Diminished immune function.

- Increased risk of blood clots.
- Increased risk of infection due to bacteria, viruses, molds, and fungi.
- Increased risk of developing autoimmune diseases.

Of the above consequences, chronic inflammation is the most common problem. Chronic inflammation can cause damage to the bones, joints, cartilage and connective tissue of the body and if left unchecked can lead to gamut of conditions from allergies, arthritis, and certain respiratory conditions such as bronchitis and sinusitis, skin conditions such as eczema and psoriasis to digestive disorders such as colitis and gastritis, and even heart disease, certain types of cancer, and multiple sclerosis.

Levy [70] reported that in the 1980s and 1990s, when Japanese and European researchers compared several enzymes for potential anti-inflammatory activity, they found that serrapeptase (Serratiopeptidase) was the most effective at controlling the body's inflammatory response.

2.4.7 Cantharidin: tumour fighting substance from blister beetle (Coleopteran)

Cantharidin is a vesicant (blistering agent) produced by beetles belonging to the order of Coleoptera and the family of Meloidae [71]. It is odorless, colorless fatty acid of the terpenoid class secreted by many species of blister beetle, produced by male beetles as defense substance and given to female beetles as a copulatory gift [72, 73]. Post-copulation behavior of the female beetle involves the protection of the eggs against any potential predators by covering it with cantharidin. There are currently more than 1500 species of cantharidin-producing beetles, commonly known as *blister beetles* or *Spanish fly*, variable in color, measure up to 2.5 cm in length and neither bite or sting [74].

Cantharidin has been available synthetically since the 1950s, topical applications of cantharidin have been used predominantly as a treatment for cutaneous warts [72, 73]. In 1962 however, marketers of cantharidin failed to produce sufficient efficacy data, resulting in the FDA revision of approval of cantharidin. In 2004, FDA accepted cantharidin, the blister-causing oil found in several families of beetles as a treatment for warts and other skin problems. Moed [74] stated despite FDA directive for the removal of cantharidin from market. Some dermatologists continue to use either proprietary or non-proprietary formulations. Recent studies in cell culture and animal models have demonstrated powerful tumor-fighting properties of cantharidin. Wang [75] cited in [74] noted that cantharidin was used historically for furuncles, piles, ulcers, venomous worms and tuberculosis scrofulderma, so [74] stated that it was used orally for abdominal cases, rabies, abortifacient and anticancer agent.

Further studies showed that cantharidin contains something known as a protein blocker. Protein blockers are used to fight infections, and appeared to attack only the infected cells. Then scientists discovered it also attacked viral cells. Now studies show that the blister beetle secretions attack hostile cells – including cancer. There are studies that indicate blister beetles might be used to "battle tumors and in chemotherapy treatments. Studies from many researchers including [76–79] suggest that the tumor growth inhibitory properties of cantharidins could be attributed, at least in part, to anti-angiogenic activity. Cantharidin and some of its derivatives inhibited the proliferation, migration, and invasion of, as well as capillary-like structure formation by cultured endothelial cells.

2.4.8 Solenopsins (Formicidae: Hymenopterans)

The word Solenopsins refers to the venom of stinging ants precisely fire ants belonging to the genus Solenopsis. The venom is water-insoluble and non proteinaceous and contains hemolytic factors that cause the release of histamine and other vasoactive amines The genus have over 200 species, in the order Hymenoptera, family Formicidae and subfamily Myrmicinae, but Fitzgerald [80] stated that two are of major medical importance: *Solenopsis richteri* (black imported ant) and *Solenopsis invicta* (red imported ant). *Solenopsis richteri* originated from eastern Argentina and Uruguay [80] while *Solenopsis invicta* is a native of the Mato Grosso region of Brazil. The venom from these fire ants is Solenopsi A (trans-2-methyl6-n-undecylpiperidine) a piperdine alkaloid with cytotoxic, hemolytic, necrotic, insecticidal, antibacterial, antifungal, anti-HIV, cardiodepressant and neurologic actions [81]. Arbiser [82] showed that solenopsin A potently inhibited the growth of SVR cells (a transformed murine endothelial cell line) by inhibiting endothelial-specific signaling. This suggested that it has anti-angiogenic activity, owing to its selective inhibition of a series of kinases involved in angiogenesis, most notably phosphatidylinositol-3-kinase (PI3K) and its downstream effector Akt (Protein kinase B (PKB), a serine/threonine-specific protein kinase that plays a key role in multiple cellular processes such as glucose metabolism, apoptosis, cell proliferation, transcription, and cell migration).

Studies from Emory and Case Western published in Scientific Reports showed elements in the fire ant venom capable of reducing skin thickening and inflammation and thus a useful treatment for psoriasis condition (a chronic immune system –based disease that manifests in abnormal patches on the skin). The toxic component of Solenopsis venom is a lipid-like molecule called ceramides. Mencarelli [83] stated that ceramides are fatty acid amides of sphingosine, which play a crucial role in homeostasis of the skin and other organs. So under certain conditions they are converted to Sphingosine-1-phosphate (S1P) which causes inflammation that is usually associated with the sting. Thus scientists have created a version of fire ant venom's solenopsin (via genetic engineering) that cannot be converted to S1P and is currently used as anti-inflammatory product.

2.4.9 Extract from Chinese black ant (Polyrhachis vicina Roger)

The Chinese black ant (*Polyrhachis vicina* Roger) according to [84] belongs to the Formicidae, Hymeoptera, Insecta in zootaxy and is widely distributed in subtropical, southeast China, India, Malaysia, Sri Lanka and Bangledesh [84]. The Chinese traditional healers were the first to use extract of black mountain ants in preparation of medicine which they call "King of herbs" even though it is an extract from black Chinese mountain ants. The ants live up high in mountains and are often found among ginseng roots. Chinese men use the tonic from the black mountain ant for treatment of impotence and fatigue. Today, modern science has proven that the extract is a rich source of energy, vitamins and minerals especially those necessary for overall sexual health. Polyrhachis is a highly concentrated source of protein (42 - 67% by mass), zinc (highest amount of zinc in all known living organisms), vitamins B-12, B-1, B-2, D, and E [85]. All of these substances are vital for health and sexual performance. The rich content of zinc is one of Polyrhachis exceptional qualities because zinc is a powerful natural antibiotic which kills many pathogens including bacteria and viruses and at the same time strengthens the immune system. Tang and Dai [84] stated that *P. vicina* had the potency to potentiate immune response in animals. Besides, the anti-inflammatory, heptaprotective, immunoregulatory and analgesic activities display by the Chinese black ant is documented by [86].

Polyrhachis vicina Roger has long been found to be very rich in ecdysterone, the growth hormone of insects, which has a strong protein anabolic effect, and so contribute to the growth promoting effect of *Polyrhachis vicina* [87]. Ecdysterone is structurally similar to androgens (male hormones such as testosterone), and many suggest its use as a safer alternative to anabolic steroids.

Black ant is also reported to contain superoxide dismutase (SOD) [85, 88]. Superoxide dismutase is a powerful antioxidant made in the liver to support oxidation for energy production. It also contains substantial amounts of the mineral selenium, used by the liver to make SOD, further increasing stress-tolerance and stamina. In addition high amounts of vitamin E in black ant increase the stress – resistance benefits.

FDA laboratory analysis showed that the black ant contains sildenafil, the active ingredient in the FDA approved prescription drug viagra, used to treat erectile dysfunction (ED). Viagra works by increasing blood flow into erectile tissue by prompting blood vessel dilation. In addition to improving blood circulation to the extremities, black mountain ants increase androgen levels (a sexual hormone) responsible for regulating sexual desire in the blood. In addition to being a potential alternative to Viagra, these ants are also being studied for their usefulness in combating cancer. The need for enhancement of sexual performance stems down from the fact that sexual performance is based on the overall health of the body. Thus sexual performance of men is weakened and fatigued by nutrient deficiencies especially a lack of protein. Such men will be too tired and quite possibly be suffering from a lack of testosterone due to nutritional deficiencies affecting the hormone levels. Natural supplements such as Polyrhachis ant (Black ant) extract has proven to be a real solution because they are not drugs and so can improve sexual performance naturally and do not present worrisome side effects unlike synthetic drugs like Viagra associated with many side effects.

2.4.10 Polybia MP1 from Brazilian social wasp (Polybia paulista): a killer of cancer cells

Michelle Roberts Health editor of online BBC news of September 1, 2015 reported on the toxin in the sting (i.e. venom of the wasp) of Brazilian social wasp (*Polybia paulista*) that kills cancer cells without harming normal cells. The cancer-targeting toxin in the wasp called MP1 (Polybia-MP1) [89]. According to a new Brazilian research published in the Biophysical Journal, a molecule called MP1 kills the cancer cells by "creating holes on their lipid membrane." These holes make molecules that cancer thrives on, to leak out. Cancer cells cannot survive without these molecules and so they die within seconds. Meanwhile, the normal cells are perfectly safe as MP1 is very selective and does not harm them.

Wang *et al.* [90] stated that Polybia MP1 is a short cationic alpha helical amphiphilic peptide that has selective toxicity toward cancer cells but no haemolytic activity. Its target selectivity according to [90] is based on the binding preference to membranes containing anionic phospholipids by electrostatic driving. These promising findings led to anti-cancer therapies involving MP1 being currently studied. According to [91, 92] cancer cell membranes are now known to lose the asymmetric transmembrane distribution of phospholipids that is observed in healthy cells. In healthy mammalian cells, the anionic aminophospholipid PS (phosphatidylserine) is predominant in the inner membrane leaflet and zwitterionic phospholipids are predominant in outer membrane leaflet. In such cells, the phospholipid asymmetry is maintained by a family of aminophospholipid translocases that catalyze the transport of PS from the outer to the inner membrane leaflets [93]. However, in apoptotic and cancer cells, PS is found to also be located in the outer monolayer of the plasma membrane in significant proportions [91, 92].

Zhao *et al* [94] demonstrated that Polybia-MPI displays potent antibacterial activity against both gram-positive and gram –negative bacteria. Wang *et al.* [95, 96] further stated that polybia-MPI has potent antifungal and antitumor activity and low toxicity to human red blood cells and normal fibroblasts.

2.4.11 Natural antibiotic from cockroach brain

The ability of bugs and other flies to survive unscathed in dirty environments with a heavy load of pathogenic microbes has attracted the interest of microbiologists and other researchers [97]. Fazackarley [98] reported the findings of Simon Lee and Naveed Khan, an Associate Professor of Molecular Microbiology presented to Society for General Microbiology's autumn meeting in Nottingham and published on the University of Nottingham website, in which the researchers described how his group identified nine different molecules in the brain tissues of the cockroach (*Periplaneta americana* (Blattidae) and desert locust (*Schistocerca gregaria*) that were toxic to bacteria. The group found that the tissues of the brain and nervous system of the insects were able to kill more than 90% of Methicillin-resistant *Staphylococcus aureus* (MRSA) and *Escherichia coli*, without harming human cells. Simon Lee stated that these molecules could eventually be developed into curative applications for the treatment of E. coli and MRSA infections that are increasingly resistant to current drugs. These new antibiotics could potentially provide alternatives to currently available drugs that may be effective but have serious and unwanted side effects. Yu *et al.* [99] and He [100] reported that pharmacological research has revealed that *P. americana* has anti-tumor effects and is able to enhance immunity, promote tissue repair, stabilize blood pressure, improve microcirculation, protect the liver, and act as an anti-inflammatory, anti-bacterial, and anti-viral agent as well as an analgesic and anti-oxidant. Dai *et al.* [101] on the other hand noted that the active ingredients isolated from *P.americana* have been developed into clinical drugs in China namely "Xiaozheng Yigan tablet", "Kangfuxin liquid", "Ganlong capsule" and "Xinmailong injection". XiaozhengYigan tablet is an oral tablet with potent anti-tumour effects and anti-bacterial activity [102]. Ou *et al.* [103] cited in [102] noted that the drug reduces liver inflammation, promotes the recovery of liver function, and reduces liver fibrosis in patients with hepatitis-B virus. *Periplaneta americana* (Blattidae), is also used for a variety of other conditions including: heartburn, asthma, stomach ache, intestinal colic, earache, alcoholism, epilepsy, vomit, boil, hemorrhage, bronchitis, diarrhea, gonorrhea, panaris, cancer, stroke, burns and menstrual cramps.

2.4.12 Other bioactive substances found in insects

Many other bioactive substances found in insects are listed in the **Table 3**.

2.5 Arthropods in Orthodox medicine

Blood-Feeding Insects and Prevention of clot formation

For long blood sucking insects have always been associated with diseases they transmit: Mosquitoes- malaria, Tse tse fly- trypanosomiasis, tick – East coast fever. The use of insect in traditional healing by practitioners of Eastern medicine in prevention of blood clot formation or thrombosis was not given much consideration. However, in recent times, some proteins in the saliva of blood sucking insects are found to possess anti-coagulation properties and these are

Bioactive substance	Insect	Properties	Author
Eumentin	Eumenine wasp (*Eumenes rubronotatus*)	Antimicrobial	[104]
Mastoparans	Vespid wasps	Antibacterial, antiviral, Antitumor,	[105–107]
Termicin and spinigerin	Termitidae *Pseudocanthotermes spiniger* Sjostedt, *Nasutitermis corniger* Motschulsky,	Antimicrobial peptides	[108]
Anticoagulants in the saliva of blood sucking insects a. Simukunin and analogues b. Tablysin-15, c. Anophelin, d. Nitrophorin-2 (or prolixin S) is a 20-kDa lipocalin	Many blood sucking insects Tablysin-15 is produced by the horsefly *Tabanus yao* Macquart, 1855 *Anopheles albimanus* C.R.G. Wiedemann, (Culicidae Kissing bug *Rhodnius prolixus* Stal, 1859	Prevent formation of blood clot	[109–111]
Termicin and spinigerin	*Pseudocanthotermes spiniger* Sjostedt, and *Nasutitermis corniger* Motschulsky,	Antimicrobial and Anti-parasites	[108]
Pierisins 1a, 1b, 2-5	*Pieris* (Pieridea) including the cabbage butterfly *P. rapae* Linnaeus	Antimicrobial and Anti-parasites	[112]
Myrmexins	*Devil tree ant (Pseudomyrmex triplarinus* Weddell) In Nigeria it is found on two main trees : Local pear (*Dacryodes edulis*) and African star apple (*Gambeya albidium*)	Extract of the venom from *Pseudomyrmex* spp. decreases pain and inflammation in patients with rheumatoid arthritis and reduces swelling	[113]
Isocarbostyril alkaloids	Texas grasshopper *Brachystola magna* Girard	Antitumor	[114]

Table 3.
Other bioactive substances found in insects.

found useful in modern or orthodox medicine. According to [115] arthropods in at least 23 different families or orders distributed between insect and arachnida feed on vertebrate blood and they are able to do this despite constraints imposed by a sophisticated array of hemostatic defenses. Their ability to do this is based on the presence of a wide range of antihemostatic molecules in their saliva including vasodilators, antiplatelet factors and anticoagulants. Leitch [116] reported that the molecule in the mosquito spit can thin blood clots. According to him, scientists are learning how to exploit the anti-clotting abilities of molecules in mosquito saliva and use them to create treatments for disorders like stroke or deep vein thrombosis. Professor Richard Payne an ARC Future Fellow in the Faculty of Science noted that this protein secreted in the saliva of mosquito precisely anopheles mosquito is called anophelin and is designed to prevent the host organism's blood from clotting, so the mosquito can access the host's blood for meal. He posited that anophelin targets and binds the central host blood coagulation enzyme thrombin and therefore prevents clotting. Further studies by scientists have shown that sulfate modification of the protein significantly improved its anticoagulant activity.

Today, the anti-coagulant activity of many more arthropods have been reported in literature [117]. Perez de Leon *et al.* [118] reported on the anticoagulant activity in the salivary glands of the insect vector *Culicoides variipennis sonorensis*. Research fellows at University of Sydney opined that by mimicking the anticlotting properties of the proteins in a mosquito's saliva, scientists can develop new drugs to treat conditions such as deep vein thrombosis or stroke.

These compounds in the saliva of blood feeding insects are capable of increasing the ease of blood feeding by preventing coagulation of platelets around the wound and provide protection against the host's immune response. Most of all, over 1280 different protein families have been associated with the saliva of blood-feeding organisms [119]. Ratcliffe *et al.* [120, 121] showed that most of them are pharmaceutical compounds including: inhibitors of platelet aggregation, ADP, arachidonic acid, thrombin, and PAF, anticoagulants, vasodilators, vasoconstrictors, antihistamines, sodium channel blockers, complement inhibitors, spore formers, inhibitors of angiogenesis, anesthetics, AMPs and microbial pattern recognition molecules. Champagne [115] stated that the vasodilators include amines, prostaglandins, peptides and proteins, platelet aggregation inhibitors include nitric oxide, prostaglandins, apyrase molecules that sequester ADP and a range of peptides and proteins, while the anticoagulants include a wide range of inhibitors that target thrombin and factor Xa, as well as proteins that disrupt the "tenase" prothrombinase and tissue factor/FIIa complexes.

2.5.1 Maggot therapy in wound treatment

Maggot therapy according to [122] is the application of live fly larvae to wounds in order to aid in wound debridement (cleaning), disinfection and/or healing. Maggot infestation of the wound of a living vertebrate host is called myiasis. Sun [123] defined maggot therapy as a type of biotherapy involving the introduction of live, disinfected maggots into the non-healing skin and soft tissue wound (s) of a human or animal for the purpose of cleaning out the necrotic (dead) tissues within a wound (debridement) and disinfection.

The effectiveness of maggot therapy is based on the proper control of the system. Such proper control of the system involves selecting the appropriate species and strain of fly (the species most commonly used is *Lucilia (Phaenicia) sericata*), disinfecting the larvae, using special dressings to maintain the larvae on the wound, and integrating quality control measures throughout the process. There are also conditions expected in the wound for the maggot to thrive: the wound must be moist, exuding with a good flow of oxygen, thus not all wound types are suitable for maggot therapy. This means that dry, open wounds of body cavities do not provide a good environment for maggots to feed. Dry wounds however can be made suitable for maggot therapy by moistening it with saline soaks, for at least 48 hours. Tian et al [124] reported that maggot therapy improves healing in chronic ulcers, diabetic foot ulcers, and venous leg ulcers. FDA in 2004 cleared maggots from common green bottle for use as a medical device in the US protocol for the treatment of non-healing necrotic skin and tissue wounds, pressure ulcers, venous stasis ulcers, neuropathic foot ulcers and non-healing traumatic or post-surgical wounds.

Even though reports of the usefulness of maggot therapy in wound healing is common in literature, patients and doctors still find maggot therapy very absurd. Sherman [122] reported that clinicians are often more disgusted with maggot dressings than are patients. Petherick *et al.* [125] study in this regard showed that approximately 25% of patients offered larval therapy for treatment of chronic leg ulcers preferred alternative treatments rather than maggot. Similarly, [28] found only 77% agreeing to leg ulcer treatment with maggots [126]. Health professionals are also skeptical about prescription of maggot therapy, but [127] is of the view the negativity is often replaced by acceptance as of the results delivered.

2.5.2 Apitherapy

Apitherapy according to [128] is the practice of using bee products such as honey, pollen, propolis, royal jelly and bee venom for disease prevention and treatment purposes. Cherbuliez [129] described it as the science and art of using honeybee products to maintain health and assist the individual in regaining health when sickness or accident interferes. The Apimondia Standing Commission for Apitherapy that works for the promotion of the use of bee products for apitherapy defined apitherapy as a medical concept, based on scientific foundations corroborating traditional knowledge, including: bee production procedures aimed at medical development. According to the archeologist Giorgog Marovfrydis, who was also a beekeeper, the first known testimonial on the use of bee products for therapeutics dates back to 2100BC which refers to a recipe written in cuneiform script on Sumerian clay plate found in Euphrates valley.

There is also biblical and Quranic evidence of God's directive on the use of honey. ***Exodus 3: 8, Exodus 33:3 Exodus 13:5, Leviticus 20:24, Numbers 13:27, Numbers 16:14, Deuteronomy 6:3, Deuteronomy 31:20, Joshua 5:6*** *all showed the description of the Promised Land as a land flowing with milk and honey. Deuteronomy 8:8 described it as a land which abounds in olive oil and in honey. Besides, the Bible has many other references on the divine directive for mankind to eat honey.*

Prov. 24:13 *"My son, eat thou honey, because it is good and honeycomb which is sweet to thy taste" (KJV)*

Prov. 25: 27 "It is not good to eat much honey, so for men to seek glory, their own glory, causes suffering and is not glory"

Matt 3:4 *and* ***Mark 1:6*** *talked of John the Baptist eating locust and wild honey.*

Koran ***Sura 16*** *stated that honey the origin and therapeutics properties of honey: "It comes from their bellies, a liquid of various colours with healing for mankind". Apart from the medicinal properties of honey, many other products are mentioned in literature: pollen, propolis, wax, royal jelly and venom.*

2.5.3 Honey

Manisha deb Mandal and Shyamapada [130] reported on the antimicrobial properties of honey. The antimicrobial property is due to its enzymatic production of hydrogen peroxide, low pH, and high content of sugar (high osmolarity). According to [131] the most common bacterium known to be inhibited by honey is *Streptococcus pyogenes*; another bacterium that is reported to be inhibited by honey is *Helicobacter pylorum* (a causative factor in ulcers). Abd-ElAal et al., [132] showed that honey has a more pronounced inhibitory effect (85.7%) on gram-negative bacteria (*Pseudomonas aeruginosa*, *Enterobacter spp*, *Klebsiella spp*). They also reported a 100% inhibition was observed in the case of gram positive methicillin-resistant *Staphylococcus aureus*.

As stated above three factors are responsible for the antimicrobial properties of honey: enzymatic activities, low pH and high content of sugar. One of the enzymes present in honey is glucose oxidase, produced by the bees' hypopharyngeal (head) glands. Upon dilution, the enzyme is activated and generates gluconic acid and hydrogen peroxide. Secondly, the high osmotic potential of honey is due to its high sugar concentration: that is osmotic effect, which can lead to the breakdown of bacterial membranes, thus inhibiting microbial growth. Honey's low pH also makes it difficult for bacteria to thrive in the medium.

Medical values of honey include its wound healing capability, oral health, treatment of pharyngitis, cough and gastrointestinal disorder, peptic ulcer, gastroesophageal reflux disease, gastroenteritis, constipation and diarrhea, liver and pancreatic diseases etc.

2.5.4 Bee pollen

Pollen, the male gametophyte of flowering plants is high energy material which is collected by insects and stored as food [133]. Bee pollen refers to the flower pollen that collects on the legs and bodies of worker bees as they collect pollen for the colony. It is rich in sugars, proteins, minerals, vitamins and fatty acids. It also includes nectar and bee saliva. The chemical composition of pollen varies depending on the flowers from which the pollen was collected. Kielisliszek [134] noted that bee pollen recently gained traction in the health community because it is loaded with nutrients, amino acids, vitamins and other 250 active substances. According to [133]), pollen has been used medically in the treatment of prostatitis, bleeding, stomach ulcers and some infectious diseases, although medical practitioners are yet to reach a compromise regarding these claims. Salles *et al* [135] stated that the Federal Ministry of Health in Germany recognizes bee pollen as medicine. Denisow et al [136] also stated that bee pollen is loaded with a wide variety of antioxidants including flavonoids, carotenoids, quercetin, kaempferol and glutathione.

2.5.5 Propolis

Propolis also called bee glue is a sticky substance used by worker bees to cement cracks in the hive. Almeida and Menezes [137] noted that the word propolis came from two Greek words: Pro (meaning "in defense of") and Polis (meaning "city"). Propolis are resinous substances, gummy in nature, collected from plants by the bees and used as cement to seal cracks or open spaces in the hive, sterilize the queen bee posture site, and to mummify insect invaders [137]. Ramos and Miranda [138] also described propolis as a honeybee product with a very complex chemical composition, made by gummy and balsamic material collected by bees from sprouts, flower-buds, trees and other vegetal-tissue resinous exudates. Although the chemical composition of propolis has been known for long. Ramos and Miranda [138] opined that correlation of propolis chemical composition with its pharmacological activities started forty years ago. Volpi [139] reported that twelve different flavonoids found in propolis namely pinocembrin(antifungal factor) , acacetin, chrysin, rutin, luteolin, kaempferol, apigenin, myricetin, catechin, naringenin, galangin and quercetin, two phenolic acids (caffeic and cinnamic acid) and a stilbene derivative called resveratrol. Besides, propolis also contain important vitamins: vitamin B1, B2, B6, C and E, as well as useful minerals Magnesium , Calcium, Potassium, Sodium, Copper, Zinc, Manganese, Iron and a few enzymes such as succinic dehydrogenase, glucose-6-phosphatase, adenosine triphosphatase and acid phosphatase [140].

Health benefits of propolis have been reported by many authors including oral health, gynecological care, oncological treatment, dermatological care and treatment for gastrointestinal disorder. Wieckiewwicz and Miernik [141] described the possible uses of propolis in treatment of various diseases of the oral cavity. Sneviranne *et al* [142] stated that the mouth environment is rich in bacterial flora, thus inhibition of microorganisms in the oral cavity could be its mode of action. However, other researchers reported that it works by inhibiting the enzyme glucosyl transferase of the bacterium *Streptococcus mutans*; a bacterium that produces lactic acid in the mouth that decays tooth enamel. Pereira [143] also showed that propolis restricts bacterial plaque development and periodontitis-causing pathogens because of its antibacterial properties. Today some companies include propolis in mouthwash, toothpaste and chewing gums as a remedy to toothache [144].

Again, owing to the fact that bacteria proliferation in the vagina, overgrowth of vaginal pathogens such as yeast like fungi, depletion of Lactobacillus spp., and elevated vaginal pH are the main causes of vaginal diseases. A study conducted by [145] on the application of 5% aqueous propolis solution resulted in an improvement in vaginal well-being. Pasupuleti [146] stated that in addition to providing antibiotic and antimycotic actions, propolis provides early symptomatic relief due to its anaesthetic properties.

On the issue of oncological treatment, Pasupuleti [147] reported that propolis has potential towards human breast cancer treatment due to its antitumor activity by inducing apoptosis on human breast cancer cells. Benguedouar et al [146] stated that galangin, a common flavonoid found in Algerian propolis, induced apoptosis and inhibited melanoma cells in vitro.

Pasupuleti [147] reported the use of propolis in dermatological products. According to them, its use in skin care products is based on its anti-allergy, ant-inflammation, anti-microbial properties and promotive action on collagen synthesis. Amad Oryan *et al* [148] showed that an important factor in impaired wound healing is biofilm formation and that propolis antimicrobial agents can reduce biofilm generation, thereby resulting in accelerated healing process. Propolis is also reported to reduce activity of free radicals (ROS) in the wound bed favoring the repair process. Pasupuleti [147] also reported that its effect on collagen metabolism by increasing both type I and type III collagens in tissues.

2.5.6 Royal jelly

Viuda-Martos et al [149] defined royal jelly as a thick and milky secretion produced by the hypopharyngeal and mandibular glands of young worker bees (*Apis mellifera*) used to feed the larvae. Literature reports showed that worker bees and queen bees start life as identical eggs laid by the parent queen, but whether the egg develops into a worker bee or a queen bee depends on the way each of them is fed. Queen bee larvae are fed with copious amounts of royal jelly and so differ in many respects from adult worker bees. This development has aroused the interest of researchers on the composition of royal jelly that has made it capable of determining the nature of a bee. Viuda-Martos *et al.* [149] also noted that royal jelly has a high content of bioactive compounds consisting of peptides such as royalisin, jelleines, aspimin and royalactina; polyphenols, principally phenolic acids, flavonoids and lipids such as10-Hydroxy-2-Decenoic acid. Ramadan and Al-Ghamdi [150] listed other bioactive compounds in royal jelly to include fatty acids, proteins, adenosine monophosphate (AMP) N1 oxide. Royalactin is the main compound in royal jelly that allows the morphological change of a larva into the queen bee [151]. It is therefore called a superfood and is the main reason for the longevity of the queen bee compared to the other bees. Strant [152] stated that royal jelly being a secretion of worker bees is of a more constant composition compared to other honey bee products. In her report, she also stated that the most interesting property of royal jelly is 10-hydroxy-2-decenoic acid and vitamins in the B complex which include thiamine, riboflavin, niacin, pyridoxine, pantothenic acid, biotin and folic acid, but little vitamin C, while vitamin A and E are absent or nearly so, vitamin D and K are probably also absent. Royal jelly was also reported to possess health promoting properties including antioxidants, anti-inflammatory, antibacterial, anti-tumour, hypocholesterlemiant, hepatoprotective, vasodilative and hypotensive [149].

The usefulness of royal jelly in wound healing has been reported by many authors. A study by researchers at Ain Shams University, Cairo in 2016 attributed the wound healing potentials of royal jelly to its ability to eradicate Methicillin Resistant *Staphylococcus aureus* infection [153] but the study did not mention involvement of any peptide. However recent research by Slovak Academy of Science found the peptide defensin-1-peptide as the factor responsible for wound healing ability of royal jelly. Defensin-1-peptide heals wounds by increasing keratinocyte migration and keratinocyte is largely responsible for epithelialization of the skin after injury [154].

Hethir Rodriguez an herbalist and a nutritionist, the Founder and President of Natural Fertility Info.com. reported on three studies to buttress the effect of royal jelly in reproductive health. First a study done in Japan and published in 2007 in the journal *Evid Based Complement Alternat Med.* which showed that Royal Jelly has the propensity to mimic human estrogen, similar to that of plant phytoestrogens. Estrogen is essential for healthy bone formation and gene expression, and is vital for a healthy menstrual cycle. The study also showed a potential for increased size of uterine cells in the rats studied.

Another study out of Kuyushu University in Japan, aimed to see if Royal Jelly could combat BPA growth-promoting effects on human breast cancer MCF-7 cells. BPA (bisphenol A) is a harmful chemical used in plastics that is a known xenoestrogen. The results of the study showed that Royal Jelly inhibits the stimulated growth of BPA on MCF-7 cells. Not only has BPA been linked to breast cancer, it has also been linked to poor egg health.

The third study reported is that of additional promising rat study of Iran, published in the *International Journal of Fertility and Sterility* which found that Royal Jelly "promotes folliculogensis [the maturing of follicles in an ovary] and increases ovarian hormones..." Thirty two female rats were divided into four groups, three control groups receiving either 100, 200 or 400 mg/ kg of body weight daily for 14 days. Rats in the Royal Jelly consuming control groups had increased levels of estradiol and progesterone, increased uterine and ovarian weights, as well as a significant increase in mature follicles and corpora lutea (more than one corpus luteum). Serum antioxidant levels were increased and nitric oxide levels decreased.

2.5.7 Bee venom

Bee venom is the poison in the bee sting that makes it very painful. It is a colourless, acidic liquid which the bees excrete through its stinger into the body of any intruder assumed to threaten its life. Barish [155] reported that an average adult can withstand 1000 sting whereas 500 stings could kill a child, but for an allergic person, one sting can cause death due to anaphylactic reaction (a life threatening allergic reaction in which blood pressure falls and airways closes). Africanized honey bees, also called killer bees, kill their victim in swarms. Research into the chemical composition of bee venom showed that it contains a complex mixture of proteins and amino acids, enzymes, sugars and lipids, polypeptides [156]. Melittin, a polypeptide compound, made up of 26 amino acids, comprising 50% of the dry weight of bee venom, has been shown to be antiviral, antibacterial and anticancer [156, 157]. Bee venom also contains the peptides: apamin and adolapin which possess anti-inflammatory and pain relieving properties.

2.5.8 Medicinal properties of termite

Medicinal properties of termite has been investigated by a number of researchers, [158] investigated the antibacterial ability of the termites mostly used by South

Indians for treating diseases associated with microorganism and found that 90% of alcohol extracts of three species of subterranean termites, namely; Odontotermes formosanus Shiraki, Microtermes obes Holmgren and Macrotermes estherae (Desneux) were effective against bacterial diseases. Nine termite species were recorded to be used for therapeutic purposes. Africa is the continent with the highest number of usage, followed by America and Asia respectively. The results showed that termites are useful for medicinal and food purposes to humanity. Alves and Alves [159] also reported that *Microcerotermes exiguus* contains antiviral properties and as such effective for treating cold, cough, hoarseness, asthma, catarrh, bronchitis, influenza, flu, whooping cough, sore throat, sinusitis and tonsillitis.

3. Conclusion

From the foregoing, the bioactive substances from arthropods are so enormous that it can no longer be overlooked or relegated to the background. Arthropods now hold the future of orthodox and traditional medicine. To me the discoveries on the medicinal, pharmaceutical and nutraceutical values of arthropods are quite indisputable, so the next line of action will be the utilization of the named arthropods in the production of drugs and other pharmaceutical as well as cosmetic products. I therefore recommend that research torchlight will be on large scale production of these insects, extraction of the bioactive ingredient and incorporation/ utilization in production of drugs especially for challenging diseases like cancer.

Conflict of interest

None.

Author details

Cordelia Ebenebe[1*], Simon Okweche[2], Oghale Okore[3], Valentine Okpoko[4], Maduabuchi Amobi[5], Joan Nneamaka Eze[6], Benedeth Ezenyilimba[1] and Michael Okonkwo[7]

1 Department of Animal Science and Technology, Nnamdi Azikiwe University, Awka, Anambra State, Nigeria

2 Department of Forestry and Wildlife Resources Management, University of Calabar, Cross River State, Nigeria

3 Department of Zoology and Environmental Biology, Michael Okpara University of Agriculture, Umuahia, Abia State, Nigeria

4 Department of Biology, Admiralty University of Nigeria, Ibusa, Delta State, Nigeria

5 Biological Sciences, Federal University of Kashere, Gombe State, Nigeria

6 Department of Agricultural Education, Federal College of Education, Asaba, Delta State, Nigeria

7 Department of Animal Science, Nnamdi Azikiwe University, Awka, Anambra State, Nigeria

*Address all correspondence to: ci.ebenebe@unizik.edu.ng

References

[1] Rádis-Baptista G. and Konno K. (2020). Arthropod Venom Components and Their Potential Usage Toxins (Basel). 2020 Feb; 12(2): 82. Doi: 10.3390/toxins1202008

[2] Peng Y., Xi W., Zhang K., Guo A. (2007). Experience improves feature extraction in drosophila J. Neurosci. 27 (19):5139-5145.

[3] Shimpi R, Chaudhari P, Deshmukh R, Devare S, Bagad Y, et al. (2016) A review:pharmacotherapeutics of bee venom. *World J Pharm Pharmaceut Sci* 5: 656-66

[4] Lee G. and Bae H. (2016). Anti-inflammatory applications of Melitinin a major component of bee venom: Detailed mechanism of action and adverse effects. Molecules 21:616

[5] Riordan F.J. (2003). Angiotensin-I-converting enzyme and its relatives volume 4, Article number: 225 (2003)

[6] Rodríguez, L. C., Faúndez, E., Seymour, J., Escobar, C. A., Espinoza, L., Petroutsa, M., Ayres, A. & Niemeyer, H. M. (2005). Factores bióticos y concentración de ácido carmínico en la cochinilla (Dactylopius coccus Costa) (Homoptera: Dactylopiidae). Agricultura Técnica, 65, 323-329. DOI:10.4067/S0365- 28072005000300001

[7] Juliane D. (2012). "Investigating Shellac: Documenting the Process, Defining the Product" (PDF). Project-Based Masters Thesis, University of Oslo. p. 27. Retrieved 13/1/2021

[8] Raman, A. (2014). Discovery of *Kerria lacca* (Insecta: Hemiptera: Coccoidea), the lac insect, in India in the late 18th century. *Current Science* 106(6), 886

[9] Das **S**, Sharon E Jacob. (2011). Shellac Dermatitis 22(4):220-2.

[10] Verheyen G.R., Ooms T., Vogels L., Vreysen S. Bovvy A., Miert S.V. and Meersman F, (2018). Insects as an Alternative Source for the Production of Fats for Cosmetics. *J Cosmet Sci.* 69(3):187-202.

[11] Ediriweera E. R. H. S. S and Premarathna N. Y. S Medicinal and cosmetic uses of Bee's Honey – A review Ayu. 2012 Apr-Jun; 33(2): 178-182. doi: 10.4103/0974-8520.105233

[12] Needham AW. Health benefits of honey. 2008. [Last cited on 2009 Oct 12].Availablefrom http://www.bees-online.com/HealthBenefitsOfHoney.htl.m

[13] Burlando B. and Cornara L. (2013). Honey in dermatology and skin care: A Review Journal of Cosmetic Dermatology. https://doi.org/10.1111/jocd.12058

[14] Kunz R. I., Brancalhão R.M.C., , Ribeiro, R.F.C. Maria Raquel Marçal Natali Silkworm Sericin: Properties and Biomedical Applications. Research International 2016(2):1-19 DOI: 10.1155/2016/8175701

[15] Voegeli R., Meier and R. Blust (1993). Sericin silk protein: Unique structure and properties. Cosmetics and Toiletries 108: 101-108.

[16] Yamada H., Yamasaki K. and I. C. Zozald (2001). Nail Cosmetics containing sericin. Patent EP 1632214 A1

[17] Padawar M.N., Pawar, A P., Daithankar A.V., and K.R. Mahadik. (2005). Silk sericin as a moisturizer: An in vivo study. Journal of Cosmetic Dermatology 4(4): 250-257.

[18] Sharma D.C. (2019). This Silk protein can help make beauty and skincare products. www.downtoearth.org.in.health. Accessed on 19th December, 2020.

[19] Dash R., Mandal M., Ghosh S.K. and Kundu S.C. (2008). Silk sericin of tropical tasar silkworm inhibits UVB-induced apoptosis in human skin keratinocytes. Molecular and Cellular Biochemistry 311 (1-2): 111-9.

[20] Wu, Q. Patočka J. and Kuča K. (2018). Insect Antimicrobial Peptides, a Mini Review Toxins (Basel) 10 (11): 461. doi: 10.3390/toxins1011046

[21] Mylonakis M.E¨, Kalli , and Rallis T S, (2016). Canine parvoviral enteritis: an update on the clinical diagnosis, treatment, and prevention Vet Med (Auckl) 11(7) :91-100. doi: 10.2147/VMRR.S80971

[22] Bulet P., Stocklin R. and L. Menin (2004). Antimicrobial Peptides from Invertebrates to Vertebrates. Immunity Rev 198:167-184

[23] Yi H., Chowdhury M., Huang Ya-Dong, Yu X. (2014). Insect Antimicrobial Peptides and Their Applications. Applied Microbiology and Biotechnology 98(13). DOI: 10.1007/s00253-014-5792-

[24] Rajkhowa D., Rokozeno M and K. Deka (2016). Insect-Based Medicines: A Review of Present Status and Prospects of Entomo-Therapeutic Resources for Human Ailment. International Journal of Agriculture Environment and Biotechnology 9(6):1069 DOI: 10.5958/2230-732X.2016.00135.2

[25] Hancock REW, Sahl HG (2006) Antimicrobial and host-defense peptides as new anti-infective therapeutic strategies. Nat Biotech 24:1551-1557

[26] Jozefiak A. and Engberg R. M. (2017). Insect proteins as a potential source of antimicrobial peptides in livestock production. A review. J. Anim. Feed Sci. 26(2):87-99

[27] Ramos –Elorduy de Concini J Pino Moreno J.M. (1988). The utilization of insects in the empirical medicine of ancient Mexicans. J. Ethnobiol 8: 195-202

[28] Kwon, S.Y., Xiao, H., Glover, B.P., Tjian, R., Wu, C., Badenhorst, P. (2008). The nucleosome remodeling factor (NURF) regulates genes involved in Drosophila innate immunity. Dev. Biol. 316(2): 538--547

[29] Dai H., Rayaprolu S. and Gong Y. (2008). Solution structure, antibacterial activity, and expression profile of Manduca sexta moricin. Journal of Peptide Science 14 (7):855-63 DOI: 10.1002/psc.1016

[30] Brown, S.E., Howard, A., Kasprzak, A.B., Gordon, K.H. and East, P.D. (2008) The 555 discovery and analysis of a diverged family of novel antifungal moricin-like 556 peptides in the wax moth Galleria mellonella. Insect Biochemistry and Molecular 557 Biology, 38, 201-212.

[31] Budnik B.A., Olsen J.V., Egorov T.A..... Galkina (2004). De novo sequencing of antimicrobial peptides isolated from the venom glands of the wolf spider Lycosa singoriensis.Journal of Mass Spectrometry 39(2)

[32] Chantawannakul (2020) Frontiers in Bioscience, Landmark, 25, 179-200,

[33] Miyoshi N., Saito T., Ohmura T., Kuroda K., Suita K., Ihara K and E. Isogai (2016). Functional Structure and AntiMicrobial Activity of Persulcstusin, An AMP from Hard Tick *Ixodes persulcatus*. Parasites and Vectors 9: 85. DOI:10.1186/s13071-016-1360-5

[34] Choi C. S., I. H. Lee, E. Kim, S. I. Kim and H. R. Kim (2000). Antibacterial properties and partial DNA sequences of cecropin-like antibacterial peptides from the common cutworm, *Spodoptera litura*. *Comp*

Biochem Physiol C Pharmacol Toxicol Endocrinol (125):287-297. DOI: 10.1016/S0742-8413(99)00117-

[35] Tene N.J. Bonnafé E., Berger F., ... Treilhou M. (2016). Biochemical and biophysical combined study of bicarinalin, an ant venom antimicrobial peptide Peptides 79. DOI: 10.1016/j.peptides.2016.04.001

[36] Chernysh S., Kim S.I., and Bekker G. (2002). Antiviral and antitumor peptides from insects. Proceedings of the National Academy of Sciences 99(20):12628- 32. DOI: 10.1073/pnas.19230189

[37] Elhag O., D. Zhou, Q. Song, A. A. Soomro, M. Cai, L. Zheng, Z. Yu and J. Zhang: Screening, expression, purification and functional characterization of novel antimicrobial peptide genes from *Hermetia illucens* (L.). *PLoS ONE* 1

[38] Zdybicka-Barabas A. and M. Cytryńska (2011). Involvement of apolipophorin III in antibacterial defense of *Galleria mellonella* larvae. *Comp Biochem Physiol B Biochem Mol Biol* 158, 90-98 DOI: 10.1016/j.cbpb.2010.10.001

[39] Herman L.L., Padala. S. A., Annamaraju P. and Bashir K (2020). Angiotensin Converting enzyme Inhibitors (ACEI) In : StatPearls (Internet). Treasure Island: StatPearls Publishing.

[40] Cito A., Botta M., Valeria F. and Dreassi E. (2017). Insects as Source of Angiotensin Converting Enzyme Inhibitory Peptides. Journal of Insects as Food and Feed 3(4):231-240 DOI: 10.3920/JIFF2017.0017

[41] Boschin G., Scigliuolo G.M., Resta D., Arnoldi A. (2014) ACE-Inhibitory Activity of Enzymatic Protein Hydrolysates from Lupin and Other Legumes. Food Chem; 145:34

[42] Kwak H.W., Ju J.E., Shin M. et al (2017). Sericin Promotes Fibroin Silk I Stabilization Across a Phase – Separation. Biomacrmolecules ISSN 1525-7797

[43] Ryu K., Lee H., and Kim I. (2002). Effects and Mechanisms of Silkworm Powder as a Blood Glucose-Lowerinly Agent. International Journal of Industrial Entomology 4(2): 93 - 100

[44] Seo C., Um I C, Rio C. W., and Kang M.Y. (2011). J. Agric. Food Chem. 59 (8): 4192-4197.

[45] Martínez-Mora C., Mrowiec A., García-Vizcaíno, E. M. and F. José Nicolás (2012)/ Fibroin and Sericin from Bombyx mori Silk Stimulate Cell Migration through Upregulation and Phosphorylation of c-Jun. PLoS ONE 7(7):e42271, DOI: 10.1371/journal.pone.0042271

[46] Aramwit P. and Sangcakul A. (2007): The Effects of Sericin on Wound Healing in Rats. Biosci. Biotechnol. Biochem 2473 -2477

[47] Ersel M., Uyanikgil.... And Cetin E. O. (2016). Effects of Silk Sericin on Incision Wound Healing in a Dorsal Skin Flap Wound Healing Rat Model. Int. Med. Sci. Monit 22: 1064-1078.

[48] Roos N. and Van Huis A. (2017). Consuming insects: Are there health benefits?. Journal of Insects as Food and Feed 3(4):225-229. DOI: 10.3920/JIFF2017.x007

[49] Felton G. W., Summers C. B. (1995). Antioxidant systems in insects. Arch Insect Biochem Physiol 29 (2):187-9 DOI: 10.1002/arch.940290208

[50] Tasaki E., Kobayashi K., Matsuura K. and Iuchi Y. (2017). An Efficient Antioxidant System in a Long-Lived Termite Queen Plos One. https://doi.org/10.1371/journal.pone.0167412

[51] EFSA (European Food Safety Authority), 2017. Webinar: rapid risk assessment tools for animal disease outbreaks. Available online: https://www.efsa.europa.eu/en/events/event/171127

[52] Burton O.T. and Zaccone P. (2007). The potential role of chitin in allergic reactions Trends in Immunology 28(10):419-22. DOI: 10.1016/j.it.2007.08.005

[53] Islam Rady, Imtiaz A. Siddiqui, Mohamad Rady, and Hasan Mukhtar (2017). Melittin, a major peptide component of bee venom, and its conjugates in cancer therapy. Cancer Lett. 402: 16-31 doi: 10.1016/j.canlet.2017.05.010

[54] Tharsnathan R. N., F. S. Kittur, 2003. Crit. Rev. Food Sci. Nutr, 43, 61-87.

[55] Barker, D. Fitzpatrick, M. P. ; Dierenfeld, E. S., (1998). Nutrient composition of selected whole invertebrates. Zoo Biology, 17 (2): 123-134.

[56] Finke, M. D., 2002. Complete nutrient composition of commercially raised invertebrates used as food for insectivores. Zoo Biology, 21 (3): 269-285.

[57] Rinaudo, M. (2006) Chitin and Chitosan: Properties and Applications. Progress in Polymer Science, 31:603-632. http://dx.doi.org/10.1016/j.progpolymsci.2006.06.001 Chitin and chitosan

[58] Dutta P. K., Dutta J, Chattopadhyaya M. C. *et al* (2004) Chitin and chitosan: novel biomaterials waiting for future developments. J. Polym. Mater. 21:321-334

[59] Shanta P., Paras Y., and Rameshwar A. (2015). Appllication of Chitin and Chitosan in Industry and Medical Science: A Review. Nepal Journal of Science and Technol 16 (1): 99- 104.

[60] Czarnetzki BM, Thiele T, Rosenbach T (1990) Evidence for leukotrienes in animal venoms. J Allergy Clin Immunol 85: 505-509. pmid:1968071

[61] Hoffman D.R. (2006). Hymenoptera venom allergens. Clin. Rev.Allergy Immunol 30:109-128

[62] Oršolić N (2012) Bee venom in cancer therapy. *Cancer Metastasis Rev* 31: 173-194.

[63] Han SM, Lee KG, Pak SC (2013) Effects of cosmetics containing purified honeybee (*Apis mellifera* L.) venom on acne vulgaris. *J Integr Med* 11: 320-326

[64] Cho SY, Shim SR, Rhee HY, Park HJ, Jung WS, et al. (2012) Effectiveness of acupuncture and bee venom acupuncture in idiopathic Parkinson's disease. *Parkinsonism Relat Disord* 18: 948-952.

[65] Shimpi R. Chaudhari P., Deshmukh R. Devare S. Bagaf Y. et al (2016). A review: Pharmacotheapeutics of bee venom. World Journal of Pharm Pharmaceut Sci. 5:656-667.

[66] Aufschnaiter A., Khalifa S., Buettner S. and ....Kohler V. (2020). Apitoxin and Its Components Against Cancer, Neurodegeneration and Rhematoid Arthritis: Limitations and Possibilities. Toxin.12(2): 66

[67] Raghuraman H1, Chattopadhyay A (2007) Melittin: a membrane-active peptide with diverse functions. *Biosci Rep* 27: 189-223

[68] Ethiraj S. and Guinath (2017). Production, purification, characterization, immobilization, and application of Serrapeptase: a review. Front. Biol. Vol. 12 (5) : 333- 348 https://doi.org/10.1007/s11515-017-1461-3

[69] Gupte V. and Luthra U. (2017). Analytical techniques for serratiopeptidase: A review J Pharm Anal. 2017 Aug; 7(4): 203-207. doi: 10.1016/j.jpha.2017.03.005

[70] Levy J. (2018). Serrapeptase: Beneficial Anti-Inflammatory Enzyme or Just Hype? https://draxe.com/nutrition/serrapeptase/

[71] Ellenhorn M. J. (1997). *Ellenhorn's Medical Toxicology: Diagnosis and Treatment of Human Poisoning.* 2nd ed. Baltimore, Md Williams & Wilkins;

[72] Al-Dawsari N. A. and Masterpol K. S. (2016). Cantharidin in Dermatology. Skinmed, 14 (2):111-114

[73] Torbeck R, Pan M., DeMoll E., Levitt J. (2014). Cantharidin: a comprehensive review of the clinical literature. Dermatol Online J. Jun 15; 20(6)

[74] Moed L., T.A. Shwayder and M.W. Chang (2001). Cantharidin Revisited: A blistering defense of an ancient medicine

[75] Wang C.C., Wu C.H., Hseh K. J. (1989). Cytotoxic effects of cantheridin on the growth of normal and carcinoma cells.Toxicology 147 (2):77-78

[76] Yu T, Hou F, Liu M, Zhou L, Li D, et al. (2012) Norcantharidin anti-angiogenesis activity possibly through an endothelial cell pathway in human colorectal cancer. *Asian Pac J Cancer Prev* 13: 499-503.

[77] Zhang L, Ji Q, Liu X, Chen X, Chen Z, et al. (2013) Norcantharidin inhibits tumor angiogenesis via blocking VEGFR2/MEK/ERK signaling pathways. *Cancer Sci* 104: 604-610.

[78] Y-Z F., Fu J-Y, Zhao Z-M, Chen C-Q. (2007) Inhibitory effect of norcantharidin on the growth of human gallbladder carcinoma GBC-SD cells in vitro. *Hepatobiliary Pancreat Dis Int* 6: 72-80.

[79] Wang H, Sun W, Zhang WZ, Ge CY, Zhang JT, et al. (2014) Inhibition of tumor vasculogenic mimicry and prolongation of host survival in highly aggressive gallbladder cancers by norcantharidin via blocking the ephrin type a receptor 2/focal adhesion kinase/paxillin signaling pathway. *PLoS One* 9: e96982

[80] Fitzgerald K.T. (2013). Insectts – Hymenoptera In Small Animal Toxicology (Third Edition). Scince direct.com

[81] Howell G, Butler J, Deshazo RD, Farley JM, Liu HL, et al. (2005) Cardiodepressant and neurologic actions of *Solenopsis invicta* (imported fire ant) venom alkaloids. *Ann. Allergy Asthma Immunol* 94: 380-386.

[82] Arbiser JL, Kau T, Konar M, Narra K, Ramchandran R, et al. (2007) Solenopsin the alkaloidal component of the fire ant (*Solenopsis invicta*), is a naturally occurring inhibitor of phosphatidylinositol-3-kinase signaling and angiogenesis. *Blood* 109:560 - 565

[83] Mencarelli C, Martinez-Martinez P. (2013). Ceramide function in the brain: when a slight tilt is enough. Cell Mol Life Sci. 70:181-203.

[84] Tang Q. and Dai Y (2016). Immunodulatory Effects of Edible Chinese black ant *Polyrhachis vicina* Roger extract using supercritical CO_2 Indian Journal of Experimental Biology 56: 411-418

[85] Shen L.R., Li D., Feng F.O. and Ren Y.C. (2006). Nutritional composition of *Polyrhachis vicina* Roger (Edible Chinese Black ant); Songlanakarin J. Sc. Technology 28:107 -114

[86] Chen J., Liu J.S., Fan C.X., Wuu F.L., Zhang M.Z., Tan I. and Zhu J.O. and

Sun W.Q. (2004). Effects of alcohol extract of *Polyrhachis vicina* Roger on Immune Function in Mice with Tumor. J. Jilin Univ. Med. Ed. 30: 543.

[87] Duan T., N Bi, and M Huang (1999). The treatment of intrauterine growth retardation with ant polyrhachis vicina Roger Zhonghua Fu Chan Ke Za Zhi;34(5):290-2

[88] Lirong S. Duo Li Feng F. and Y. Ren (2006). Nutritional composition of *Polyrhachis vicina* Roger (Edible Chinese Black ant); Songlanakarin J. Sc. Technology 28:107 -114

[89] Reid C. (2015). Scientists finally solved the mystery of why this wasp's venom can target and kill cancer cells IFL Science

[90] Wang K, Yan J, Zhang B, Song J, Jia J, Jia P, Wnag R (2009) Novel mode of action of polybia-MP1 a novel antimicrobial peptide in multidrug resistant leukemic cells. Cancer Lett 278:65-72

[91] Zwaal R.F.A., Comfurius P. and Bevers E.M. (2005). Surface exposure of Phosphatidylserine in pathological cells. *Cell. Mol. Life Sci*. 2005; 62: 971-988

[92] Schweizer F. (2009). Cationic amphiphilic peptides with cancer-selective toxicity. *Eur. J. Pharmacol.* 2009; 625: 190-19

[93] Stafford J. H. and Thorpe P. E. (2011). Increased Exposure of Phosphatidylethanolamine on the Surface of Tumor Vascular Endothelium. Neoplasia. 2011 Apr; 13(4): 299 308.doi: 10.1593/neo.101366

[94] Zhao Y., Yang A., Tu P. et al. (2017). Anti-tumour effects of the American Cockroach, *Perplanata americana*. Chin Med 12 (26). https://doi.org/10.1186/s13020.017-0149-6

[95] Wang K.R. Zang B.Z., Zang W., Yan J.X., Li J. and Wang R., Anti tumor effects , cell selectivity and structure-activity relationship of a novel antimicrobial peptide polybia MP1. Peptides 29 (6): 963-8. Doi: 10.1016/j.peptides.2008.01.015

[96] Wang F.X., N. Wu, J. T. Wei, J. Liu, J. Zhao, A. G. Ji and X. K. Lin: A novel protein from *Eupolyphaga sinensis* inhibits adhesion, migration, and invasion of human lung cancer A549 cells. *Biochem Cell Biol* 91, 244-251 (2013)

[97] Lee, S. Siddiqui, R. and Khan N.A. (2012). Animals living in polluted environments are potential sources of antimicrobials against infectious agents Pathog Glob Health 106(4): 218-223. doi: 10.1179/2047773212Y.0000000033

[98] Fazackarley J. (2010). Cockroach Brains could be a new source of Antibiotics http://www.digitaljournal.com/article/297163#ixzz6g8Kx9osI

[99] Yu S., Zhang H, Zhang T, Liu J. (2016). Research Advances in Pharmacological action and Clinical Applications in Chemical Constituents and Pharmacological Effect of *Periplanata americana*. J. Liaoning Coll. Tradit Chin Med. 18: 228-230

[100] He Z, Peng F., Song L., Wang X., Hu M., Zhao Y. and Liu G. (2007). Advances in Chemical Constituents and Pharmacological effects of Periplanata Americana. Chin J. Chin Mater Med

[101] Dai Y., Zeng M. and Xiang P. (2005). The medicinal value of cockroaches. Chin Med Mat. 28:848-849

[102] Zhao et al 2017

[103] Ou W., Zhu C. and Lin W. (1995). Observation of the Clinical Curative Effects of Xiao Zheng Yi gan Pian on Chronic Hepatitis B. J. Integr Tradit West Med Liver Dis . 5:12-13

[104] Konno K. et al (2006). Eumenitin, A Novel Antimicrobial Peptide From the Venom of the Solitary Eumenine Wasp Eumenes rubronotatus. Peptides 2006

[105] Sample CJ, Hudak KE, Barefoot BE, Koci MD, Wanyonyi MS, et al. (2013) A mastoparan-derived peptide has broad-spectrum antiviral activity against enveloped viruses. *Peptides* 48: 96-105

[106] Henriksen JR, Etzerodt T, Gjetting T, Andresen TL (2014) Side chain hydrophobicity modulates therapeutic activity and membrane selectivity of antimicrobial peptide mastoparan-X. *PLoS One* 9: e9100

[107] Wu S., M. Lu and S. Wang (2017): Antiageing activities of water-soluble chitosan from *Clanis bilineata* larvae. *Int J Biol Macromol* 102, 376-379 DOI: 10.1016/j.ijbiomac.2017.04.03

[108] Landon C, Meudal H, Boulanger N, Bulet P, Vovelle F (2006) Solution structures of stomoxyn and spinigerin, two insect antimicrobial peptides with an alpha-helical conformation. *Biopolymers* 81: 92-103

[109] Andersen JF, Pham VM, Meng Z, Champagne DE, Ribeiro JM (2009) Insight into the sialome of the black fly, *Simulium vittatum*. *J Proteome Res* 8: 1474-1488 .

[110] Tsujimoto H, Kotsyfakis M, Francischetti IMB, Eum JH, Strand MR, et al. (2012). Simukunin from the salivary glands of the black fly *Simulium vittatum* inhibits enzymes that regulate clotting and inflammatory responses. *PLoS ONE* 7: e29964

[111] Ronca R, Kotsyfakis M, Lombardo F, Rizzo C, Currà C, et al. (2012) The *Anopheles gambiae* cE5, a tight- and fast-binding thrombin inhibitor with post-transcriptionally regulated salivary-restricted expression. *Insect Biochem Mol Biol* 42: 610-620

[112] Subarayan S, Marimuthu SK, Nachimuthu SK, Zhang W, Subramanian S (2016) Characterization and cytotoxic activity of apoptosis-inducing pierisin-5 protein from white cabbage butterfly. *Int J Biol Macromol* 87: 16-27

[113] Pan J, Hink WF (2000) Isolation and characterization of myrmexins, six isoforms of venom proteins with anti-inflammatory activity from the tropical ant, *Pseudomyrme triplarinus*. *Toxicon* 38

[114] Pettit GR, Meng Y, Herald DL, Knight JC, Day JF (2005) Antineoplastic agents The Texas grasshopper. *Brachystola magna*. *J Nat Prod* 68: 1256-1258

[115] Champagne D. (2005). Antihemostatic Strategies of Blood Feeding Arthropods.Cardiovascular and Haematological Disorders 4 (4): 375- 396

[116] Leitch C. (2018). Molecule in Mosquito Spit Can Thin Blood Clots. Cell and Molecular Biology. Labroots.com

[117] Stark K. R., James A. A. (1996). Anticoagulants in vector arthropods. Parasitol Today 12(11):430-7. doi: 10.1016/0169-4758(96)10064-8.

[118] Perez de Leon A. A., Valenzuela J. G., Tabachnick W. J. (1998). Anticoagulant activity in salivary glands of the insect vector *Culicoides variipennis sonorensis* byan inhibitor of factor Xa *Exp. Parasitol.* 88: 121 – 130

[119] Ribeiro, J.M.C., Mans B.J., Arcà B. (2010). An insight into the sialome of blood- feeding Nematocera Insect Biochemistry and Molecular Biology 40(11):767-784 DOI: 10.1016/j. ibmb.2010.08.002

[120] Ratcliffe, N., Azambuja, P., and Mello, C. B. (2014). Recent advances

in developing insect natural products as potential modern day medicines. *Evid. Based Complement Alternat. Med.* 2014:904958. doi: 10.1155/2014/904958

[121] Francischetti, I. M. B., Pham, V. M., Mans, B. J., Andersen, J. F., Mather, T. N., Lane, R. S., et al. (2005). The transcriptome of the salivary glands of the female western black-legged tick *Ixodes pacificus* (Acari: Ixodidae). *Insect Biochem. Mol. Biol.* 35, 1142-1161. doi: 10.1016/j.ibmb.2005.05.007 122. \

[122] Sherman R.A. (2009). Maggot therapy takes us back to the future of wound care.New and improved maggot therapy of the 21st century. Journal of Diabetes Science and Technology 3(2):336-344.

[123] Sun, X., Jiang K. and Jingan et al (2014). A systematic Review of maggot debridement therapy for chronically infected wounds and ulcers. International Journal of Infectious Diseases 25:32-37.\Tian, X; Liang, XM

[124] Song, GM; et al. (2013). "Maggot debridement therapy for the treatment of diabetic foot ulcers: a meta-analysis". *Journal of Wound Care. 22 (9): 4629. doi:10.12968/jowc.2013.22.9.462. PMID 24005780*

[125] Petherick, J.C.; Holroyd, R.G.; Doogan, V.J. et al (2007). Productivity, carcass and meat quality of lot fed *Bos indicus* cross steers grouped according to temperament. Australian Journal of Experimental Agriculture, .42, p.389-398.

[126] Spilsbury K., Cullum N.A. and Dumville (2008). Exploring Patients Perception of Larval therapy As a Potential Treatment for Venous Leg Ulceration. Journal of Participation in Health Care and Health Policy 11(2): 148-159.

[127] Sherman R.A. (2009). "Medicine, Insects in" in Encyclopedia of Insects (Second Edition)

[128] Trumbeckaite S. et al (2015). Knowledge, Attitudes and Usage of Apitherapy for Disease Prevention and Treatment Among Undergraduate Pharmacy Students in Lithuania. Evid Based Complement Alternative Med. Doi: 10.1155/2015/172502 \

[129] Cherbuliez T. Apitherapy–the use of honeybee products. Biotherapy-History, Principles and Practice: Springer; 2013. p. 113-46.

[130] Manisha Deb Mandal and Shyamapada Mandal (2011). Honey: its medicinal property and antibacterial activity Asian Pac J Trop Biomed.; 1(2): 154-160. doi: 10.1016/S2221-1691(11)60016-6

[131] Almasaudi S.B., Alea A.M., Elsajed M., Yousef A.Q., Esan A. and Mohamad Q. (2017). Antimicrobial effect of different types of honey on *Staphylococcus aureus*. Saudi Journal of Biological Sciences, 24:1255 – 1261.

[132] Abd- EIAal A.M. El-Haddy M.R El- Mashad N.B. and A.H. El-Sebaie (2007). Antimicrobial Effect of Bee Honey in Comparison to Antibiotics on Organisms Isolated from Infected Burns. Annals of Burns Fire Disaster 20 (2): 83-88.

[133] Linskens H. F. and W. Jorde (1997). Pollen as food and medicine: A review. *Economic Botany* volume 51, Article number: 78

[134] Kieliszek M. Piwowarek K. Anna M. Stanisław K. Błażejak Anna Chlebowska- ŚmigielIwona Wolska (2018). Pollen and bee bread as new health-oriented products: A review. Trends in Food Science & Technology Volume 71: 170-180

[135] Salles J, Cardinault N., Patrac V, Berry A, Giraudet C, Collin M, Chanet A., Tagliaferri C, Philippe Denis P, Pouyet C, Boirie Y., and Walrand S. (2014). Bee Pollen Improves

Muscle Protein and Energy Metabolism in Malnourished Old Rats through Interfering with the Mtor Signaling Pathway and Mitochondrial Activity Nutrients.; 6(12): 5500-5516. doi: 10.3390/nu6125500

[136] Denisow B. et al (2016). Biological and therapeutic properties of bee pollen: A Review. Journal Sci Food and Agric. 96 (13):4303-9. Doi.10.1002/jsfa7729.

[137] Almeida E. C. de and Menezes H. (2002). Anti-Inflammatory Activity of Propolis Extracts: A Review J. Venom. Anim. Toxins vol.8 no.2 Botuca https://doi.org/10.1590/S0104 79302002000200002

[138] Ramos A. F. N.; Miranda J. L. (2007). Propolis: a review of its anti-inflammatory and healing actions. J. Venom. Anim. Toxins incl. Trop. Dis vol.13 no.4 Botucatu 2007 https://doi.org/10.1590/S1678-91992007000400002

[139] Volpi N. (2004). Separation of flavonoids and phenolic acids from propolis by capillary zone electrophoresis. Electrophoresis 25(12):1872-8 DOI: 10.1002/elps.200405949

[140] Lotfy M. (2006). Biological Activity of Bee Propolis in Health and Disease. Asian Pac J Cancer Prev, 7: 22-31

[141] Więckiewicz W., Miernik M., Więckiewicz M, and Morawiec T (2013). Does propolis help to maintain oral health? Evidence-based Complementary and Alternative Medicine Ecam, :351062 DOI: 10.1155/2013/351062

[142] Sneviratne, R. W.; Beltranena, E.; Goonewardene, L. A.; Zijlstra, R. T., (2011). Effect of crude glycerol combined with solvent-extracted or expeller-pressed canola meal on growth performance and diet nutrient digestibility of weaned pigs. J. Anim. Sci., 170 (1-2): 105-110.

[143] Pereira E.M. R. (2011). Clinical evidence of the efficacy of a mouthwash containing propolis for the control of plaque and gingivitis a phase II study.Hindawi Publishing Corporation; 11: 1-6

[144] Bertolini F. R., O. Biondi Filho, A. Pomilio, S. L. Pinheiro, and M. S. D. Carvalho, (2012). "Antimicrobial capacity of *Aloe vera* and propolis dentifrice against Streptococcus mutans strains in toothbrushes: an in vitro study," *Journal of Applied Oral Science*, vol. 20, no. 1, pp. 32-37

[145] Imhof, M., M. Lipovac, C. Kurz, J. Barta, H. C. Verhoeven, and J. C. Huber (2005). "Propolis solution for the treatment of chronic vaginitis," *International Journal of Gynaecology and Obstetrics*, vol. 89, no. 2, pp. 127-132.

[146] Benguedour L. (2016).Algerian Ethanolic Extract of Propolis and gangalin decreased melanoma tumor progression in C57BL6 mice Annales de Dermatologie et Venereologie. France, Elsevier. Pp. 294

[147] Pasupuleti V. R., Sammugam L., Ramesh, N. and Ga S.H. (2017). Honey, Propolis and Royal Jelly: A Comprehensive Review of Their Biological Actions and Health Benefits. https://doi.org/10.1155/2017/1259510

[148] Ahmad Oryan et al (2018). Potential role of propolis in wound healing: Biological properties and therapeutic activities. Biomed Pharmacother

[149] Viuda-Martos M., Perez-Alvarez J. A. and Fernadez-Lopez J. (2017). Royal JellyL Health Benefits and Uses in Medicine In Book, Bee Products: Chemical and Biological Properties. Doi: 10.1007/978.3.31959689-1_1

[150] Ramadan M. F. and A. Al-Ghamdi, (2012). "Bioactive compounds and health-promoting properties of royal jelly: a review," *Journal of Functional Foods*, vol. 4, no. 1, pp. 39– 52.

[151] Kamakura M. (2011). Royalactin induces queen differentiation in honeybees. Nature 473 (7348): 478-483. Doi:101038/Nature10093 (PubMed).

[152] Strant M., Yucel B., Topal E. [...] and Alina V. (2019). Use of Royal Jelly as Functional Food in Human and Animal Health. Hayvansai Uretim 60(2):131-144.

[153] Mona H.. El-Gayar... Hassouna N.A. Antivirulence and wound healing effects of royal jelly and garlic extract for the control of MRSA skin infections. Wound Medicine 13:18- 27

[154] Bucekova M. et al (2017). Bee-derived Antibacterial Peptide, Defensin -1, PromotesWound Reepthelialisationin Vivo and in Vitro. Scientific Reports Vol 7. Article No 7340

[155] Barish R. A. (2020). Bee, Wasp, Hornet, and Ant Stings. https://www.msdmanuals.com/home Br J Pharmacol. 2018 Dec; 175(23): 4310-4324

[156] Hyun-Jin An, Jung-Yeon Kim, Woon-Hae Kim, Mi-Gyeong Gwon et al (2018)Therapeutic effects of bee venom and its major component, melittin, on atopic dermatitis *in vivo* and *in vitro* doi: 10.1111/bph.14487

[157] Huerta-Reyes M., Anselme C., Cherqui A. and Decoca G. (2017). Exploration Through the Venoms From Hymenoptera as Potent Therapeutic Agents in Cancer Therapy. Int. J. of Pharmacology ol 13 (6): 507-515.

[158] Solavan A, Paulmurugan R, Wilsanand V. (2007). Antibacterial activity of subterraneantermites used in South Indian folk medicine. Indian Journal of Traditional Knowledge 6: 559-562

[159] Alves, R.R.N. and Alves, H.N. (2011). The faunal drugstore: Animal-based remedies used in traditional medicines in Latin America. J Ethnobiol Ethnomed; 7:1-4

Chapter 2

Pollinators: Their Evolution, Ecology, Management, and Conservation

Victoria Wojcik

Abstract

Insect pollinators are a rich and diverse group of species that have coevolved with plants to create biodiverse and productive landscapes that support ecosystem services. Bees, beetles, flies, butterflies, moths, and even ants participating in moving pollen within and between flowers, assisting the reproduction of more than 80% of all flowering plants. The value of insect pollinators to ecosystems and economies is both large and immeasurable. One of three bites of food eaten is pollinated, and countless raw materials and natural products are the result of the visitation of flowers by insects. Yet, these keystone species face survival challenges driven by habitat loss, pests, disease, pesticides, and climate change. Conservation, restoration, and management seek to build back resilience into these systems, without which our world would be unrecognizable.

Keywords: pollination, coevolution, ecosystem services, Hymenoptera, Lepidoptera, Diptera, Coleoptera, agriculture, conservation

1. Introduction

Pollination is the movement of pollen from the anther (male part) to the stigma (female part) of a flower. It is the manner in which seed plants, or *spermatophytes,* reproduce. Although this process can occur by wind or water, the dominate form in which these sessile organisms move their gametes is by animal vector. Animal pollination is a foundational ecosystem service, with an estimated 80% of flowering plants requiring or benefitting from pollination [1]. These intricate systems evolved over 140 million years ago and are the reason for the rich, diverse landscapes we enjoy and benefit from today. A diversity of animals provide pollinator services, but the majority of pollinators are insects; the most effective and important of these anthophilic insects are bees (Hymenoptera, Apoidea). These vital insects face many challenges that threaten their survival, the majority of which are the direct results of human impacts on the environment. Management in agricultural systems aims to promote bee health in order to preserve crop yields. Conservation in natural areas seeks to maintain and ensure that insect pollinator populations continue to support our ecosystems and livelihoods.

2. Evolution of pollination

2.1 Origins and the fossil record

Animal-mediated pollination evolved some 140 million years ago, when plants developed floral rewards in the form of nectar and pollen, along with attractive floral displays. Flowers evolved to attract animal visitors that would then spread pollen, in a targeted way, from one flower to another. The earliest pollinators were *palynivores*, feeding on pollen, and in the process of visiting flowers for food, got contaminated with pollen grains, moving them from one flower to the next. Beetles are seen emerging in the fossil record in the Jurassic, before the emergence of flowers, and are considered to be the first pollinators. Fossil records document beetle pollination in the Cretaceous, with the earliest record of pollen on beetles see in amber-preserved *Cretoparacucujus cycadophilus* carrying cycad pollen dated at 100 million years old [2]. Present-day beetle pollination systems of magnolias resemble these early origins. In the discussion of the earliest pollinators there is fossil evidence from Spain that shows thrips (Thysanoptera) preserved in amber that are covered in gains of *Cycadopites* pollen [3], suggesting a diversity of prehistoric insects had associations with flowers.

Bees evolved somewhere between 140 to 70 million years ago from wasps that switched their feeding habits from a carnivorous diet of mostly other insects to pollen and nectar. The oldest fossil record of bees is a specimen of stingless bee, *Trigona prisca*, found in the United States that is dated to between 96 and 74 million years old, from the Upper Cretaceous [4]. Fossil records of *Apis* bees have been found in Western Germany that date to the Lower Miocene (approximately 25 million years ago). These are considered to be precursors to the modern day honey bee, *Apis mellifera*, arguably the most important insect pollinator. Flies appear in the fossil record about 200 million years ago; moths and butterflies are markedly older, appearing as early as 400 million years ago.

2.2 Coevolution

The coevolution of plants and insect pollinators has resulted in mutualisms that are an incredible diversification of forms and functions as plants developed ways to ensure visitation, fidelity, and pollen transfer to secure their reproduction. From the plant's perspective, attracting a visitor that will coincidentally get covered in pollen before visiting another food source was the first step in the development of tailored systems that allowed plants to produce fewer pollen grains. Flowers evolved as specialized structures to further attract insect visitors. This included elaborate shapes, scents, and colours. Scents and colours advertise flowers to pollinators looking for food in new landscapes.

The size and shape of the flower facilitates the transfer of pollen to the pollinator, with adaptations to each group. Complicated morphologies that squeeze, and in some chases trap, pollinators work to increase the chance of transferring pollen. Orchids are famous for specialized pollination systems, including trap pollination, as exhibited by *Cypripedium* species. Orchid bees attempting to visit Lady's Slippers flowers slip into a chamber where they are trapped temporarily, and presented with a single, narrow opening through which to exit. As they squeeze through this narrow opening they move past the sticky *pollinia* (large pollen sacks) that are deposited on their backs. Another group of flowers with trap blossoms are the *Arums*. These plants trap pollinating flies within their base where the anthers are found until the pollen is mature, opening to release pollen-covered flies and beetles that will then go and visit others flowers. Many trap-flowers use deception to attract insect visitors, mimicking reproductive pheromones or the scent of food sources (i.e., carrion). Another form of deceptive pollination used by some flowers, such as orchids, is Pouyannian mimicry: tricking a male visitor into thinking there is a receptive mate. The mirror orchid, *Ophyrs speculum*, appears in colour and shape like the female scoliid wasp (Scoliidae), this tricks the male into attempting to mate, referred to as *pseudocopultion*, which then triggers the release of pollen onto the unsuspecting insect.

Concurrently insects evolved individualized morphologies and behaviours to increase their fitness and optimize their ability to capitalize on floral resources. Insects are likely pollen vectors, often possessing hairs or scales that pollen attaches to easily. Bees, the most effective and efficient pollinators, have branched hairs that often occur in dense aggregations called *scopa* that are specialized to carry pollen. Unlike other pollinators that visit flowers to feed on nectar and pollen, accidentally getting contaminated with pollen, bees actively collect pollen to provision for their offspring. Early bees collected pollen in their digestive systems, holding it in their *crop*. Some more primitive bees in the family Colletidae still make use of ingesting pollen as they do not have the pollen-carrying hairs on their body. Most bees, however, carry pollen in an external load, which can also be used as an identifying characteristic.

3. Pollination systems

Over the last 120 million years, pollinators have diversified an incredible amount. Bees, butterflies, moths, flies, beetles, and wasps have evolved to fill specific niches in the environment. The generalized characteristics of plant-pollinator systems are commonly described in pollination syndromes; predictive sets of plant morphology and phenology that align with the preferences of pollinators. Insect pollinator syndromes include bee and wasp pollination (melittophily); butterfly pollination (psychophily); moth pollination (phalaenophily); fly pollination (myophily and sapromyophily); beetle pollination (cantharophily); and even ant pollination (myrmecophily) (**Table 1**). Using the theory of pollination syndromes, Charles Darwin postulated that the pale and fragrant Christmas Orchid, *Angraecum sesquipedale*, was likely pollinated by a moth with a proboscis equal in length to the long nectary possessed by this flower, measuring more than 25 cm in some cases. Darwin passed away before he ever validated his prediction, but two decades later a sphinx moth, *Xanthopan morganii,* with in impressive 30 cm proboscis was found to in fact be a visitor of the orchid.

3.1 Pollinator syndromes

3.1.1 Canatharophily

Beetle pollination: With more than 380,000 species of beetles globally it is assumed that more than 90% of all pollinated plants have associations with beetles. Beetles may not be the primary pollinator, but their ubiquitous presence makes them important anthophiles. Flowers that are pollinated by beetles, or those that present characteristics to attract beetles, are commonly large; pale white, cream or green in colour; and have a heavy scent that is sometimes foul, mimicking carrion. Beetle bodies are robust and they are often clumsy, correspondingly beetle-pollinated flowers are often flat or disc-shaped. The pollen is easily accessible as beetles have minimal ability to handle and manipulate it. Beetles have been called destructive pollinators, visiting flowers to feed on floral

Floral Trait	Colour	Shape	Nectar Guides	Odour	Pollen	Nectar	Bloom
Canatharophily (beetle)	pale/dull, white, green	large, bowl-shaped	none	strong fruity or foul	abundant	accessible nectaries	day
Melittophily (bees, pollen wasps)	yellow, blue, purple, ultraviolet	complex, disc, or tubular	present, obvious	fragrant, pleasant	abundant	abundant	day
Psychophily (butterflies)	red, pink, purple, bright	disc shaped, landing pad	present	strong and fragrant	limited	abundant, deep	day
Phalaenophily (moths)	pale, cream, white, yellow	tubular, with and without landing pad	none	strong and fragrant	limited	abundant, deep	day and night
Myophily (flies)	yellow, white	disc shaped	present	strong and fragrant	limited	abundant	day and night
Sapromyophily (carrion flies)	marron, green, dull	funnel or trap, small	none	strong and putrid	limited	absent	day and night
Myrmacophily (ants)	varied	open, small	none	none	varied	abundant, extra-floral nectaries	day

Table 1.
Insect Pollinator Syndromes, various floral characteristics that correspond to generalized preferences and use patterns of seven types of insect visitors.

parts, or the pollen directly. For this reason, many beetle-pollinated plants have extra protective structures around their ovaries. Pollen is the dominant reward produced by beetle-pollinated flowers.

3.1.2 Melittophily

Bee pollination: Flowers that are visited by bees have perhaps the greatest diversity in form. Bees often make use of open, disc-shape flowers, but are also able to access more complicated floral forms, even those that require complicated floral handling. For example, flowers in the pea family (Fabaceae) have a complex, asymmetrical morphology, with a keel petal that covers the opening and access to nectar and pollen. Both strength and persistence are required to access the floral rewards in such blossoms. Bee-pollinated flowers are commonly yellow, blue, or purple (**Figure 1**). They also commonly possess ultraviolet colouration as bees can see this part of the light spectrum. Many bee-pollinated flowers have nectar guides, areas of colour that highlight the location where they nectaries are located. Nectar guides focus pollinator effort and optimize pollination. Bee-pollinated flowers are heavily scented, allowing an additional chemical sensory way for bees to locate them. Although melittophily predominantly refers to bees, pollen wasps are attracted to

Figure 1.
Small sweat bee (Halictus sp.) visiting a compound, disc-shaped, purple flower (Asteraceae) representative of melittophily. [photo credit: Amber Barnes].

the same set of morphological characteristics. Bees are specifically adapted to be the ideal pollinators. While most pollinating insects visit flowers to feed on nectar, accidentally getting covered in pollen, bees purposely collect pollen to feed their young. Nectar is a food source for adult bees, and this high-energy carbohydrate food powers their metabolically expensive flight. For this reason, bee-pollinated flowers provide abundant nectar and pollen.

3.1.3 Psychophily

Butterfly pollination: Flowers pollinated by butterflies are commonly disc-shaped or compound in structure to provide an easy landing pad. Members of the family Asteraceae are widely visited by butterflies. Butterfly-pollinated flowers are obvious and showy, usually pink, red, or purple, and they are highly scented. Unlike bees, butterflies can see the colour red, and correspondingly many of the flowers visited by butterflies are red. Butterflies feed on nectar using a proboscis and are looking for abundant nectar. Nectaries can be located in tubes or spurs that are accessible by this long proboscis. The anthers of butterfly-pollinated flowers are

Figure 2.
An Eastern Tiger Swallowtail butterfly (Papilio glaucus) feeding on a pink flower (Asteraceae) with a large disc-shaped landing pad representative of psychophily. [photo credit: Amber Barnes].

usually located in places that would deposit pollen on the heads or undersides of butterflies. It is common for butterflies to land on compound flowers and proceed to feed at multiple sites (**Figure 2**). Butterflies are sun-seekers that exhibit basking behaviours, enjoying flat flowers in sunny locations. A common group of specialized butterfly-pollinated plants are the milkweeds (family Apocynaceae, genus *Asclepias*). The flowers of these plants present the characteristics of psychophily: they are colourful, clustered, and tubular. Additionally, monarch butterflies, *Dannus plexippus*, are co-evolved with *Asclepias* species; the caterpillars are able to eat their toxic foliage, which contains cardenolides, resulting in the development a chemical defense against predators (**Figures 3–6**).

3.1.4 Phalaenophily

Moth pollination: Flowers pollinated by moths share some characteristics with butterfly-pollinated flowers, including nectaries in deep spurs and attractive scents. Most moths are active in the evening, or at dawn and dusk (crepuscular), so their flowers are generally less colourful and more perfumed. Moth-pollinated flowers

Figure 3.
Monarch (Danus plexippus) egg laid on the underside of a milkweed (Asclepias) leaf. [photo credit: Amber Barnes].

Figure 4.
Monarch (Danus plexippus) caterpillar feeding on milkweed (Asclepias); this behaviour instills a chemical defense in the caterpillar and butterfly. [photo credit: Amber Barnes].

tend to be white or cream coloured. The amount of nectar produced is relatively high as many moths hover while they access their food and have a high metabolic demand.

3.1.5 Myophily

Fly pollination: Pollinating flies that feed on nectar are attracted to a similar variety of floral types as bees are attracted to. Fly-pollinated flowers tend to display a diversity of colours and shapes, although they tend not to be as complex as those visited by bees. The scent emitted by these flowers is also usually quite fragrant and potent to aid flies in locating the flowers. Flies are nectar feeders, and fly pollinated flowers provide ample nectar.

3.1.6 Sapromyophily

Fly pollination by carrion or dung seekers: Less common than myophily, flowers pollinated by flies seeking animal flesh or dung as egg-laying substrates mimic these elements. They are heavily scented to mimic foul odours. The flowers can be small, colourless, or a dull purple/pink, with a sticky secretion. Trap-blossoms are

Figure 5.
Monarch (Danus plexippus) chrysalis. [photo credit: Amber Barnes].

also common. These flowers do not offer rewards; instead they trick the insect into visitation. A wonderful example of sapromyophily is the giant corpse plant, or titan arum, *Amorphophallus titanum*. The gigantic flower produced buy this arum can be upwards of 2 meters tall, and produces a strong, foul odour that resembles rotting flesh. Flesh flies and carrion beetles are attracted to this plant; as they attempt to find a meal or lay eggs they are trapped inside the giant flower until it wilts, releasing them after they have been covered in pollen.

3.1.7 Myrmecophily

Ant pollination: Ants are minor pollinators, and are not attracted to showy floral structures. In many cases the flowers of ant pollinated plants are small, often occurring in clumps. The flowers can be also located at the base of stems. The plants themselves are low-growing and accessible from the ground. Ants feed on nectar found inside of the flowers as well as on extra-floral nectaries. Ant-pollinated plants occur in arid or alpine environments. Examples of ant-pollinated plants include Small's stonecrop (*Diamorpha smallii*), alpine nailwort (*Paronychia pulvinata*), and Cascade knotweed (*Polygonum cascadense*), as well as two alpine orchid species, *Chamorchis alpine* and *Dactylorhiza viridis*.

Figure 6.
Adult monarch butterfly (Danus plexippus) feeding on the flowers of common milkweed (Asclepias speciosa). [photo credit: Amber Barnes].

3.2 Adaptations for pollination and flower visitation

3.2.1 Physical adaptations

Each arthropod pollinator group has a specific physical adaptation for accessing pollen and nectar. Most insects adapted to pollen collection are covered in hairs or scales. Bees are the most highly adapted to collecting and carrying pollen. Their bodies are covered in dense, bifurcated hairs that garb onto pollen. Many bees have dense aggregations of pollen-carrying hairs, called scopa, on their hind legs or the underside of their abdomen. The placement of pollen-carrying hairs can aid in general identification of bees. Members of the family Megachilidae (*Megachile*, *Osmia*, *Anthidium*, *Hoplitis*, and *Chalicodoma*), commonly known as leafcutter bees, mason bees, and carder bees, are all identifiable by their ventral abdominal scopa. Bees in the family Andrenidae carry pollen on the upper portion of their hind legs, while members of the family Halictidae carry pollen down the length of their hind legs. Members of the family Apidae also carry pollen on their hind legs, but generally in a more compacted pattern; some have further specialized structures called *corbicula* where they can carry a tightly packed ball (**Figure 7**); this is common in honey bees (*Apis mellifera*), bumble bees (*Bombus* spp.), and carpenter bees (*Xylocopa* spp.).

The adaptations of other insect pollinators are specific to accessing nectar and pollen for food. Flies do not actively collect pollen, but are effective pollinators because they are covered in hair. Flies feed on nectar and access it through the use of a shortened set of sucking mouth parts. Nectar taken in this manner has to be shallow and accessible. Butterflies and moths are covered in scales, not only on their wings, but also on their face, thorax, abdomen, and even their legs. Pollen is transferred to these scales when they feed, and then transferred to another flower when they make another visit. Butterflies and moths feed on nectar accessed through a proboscis, which in some cases is extremely long in order to access nectar in deep nectaries. Beetles feed on pollen, collecting it with their mouthparts. Some have additional adaptations such as rows of dense hairs on their maxilla or labium, which act as a pollen broom that helps convey pollen into their mouth.

Pollen and nectar are the dominant rewards sought by anthophiles, but some also collect various oils, resins, and other substances. Bees in the genera *Eulema*

Figure 7.
A bumble bee (Bombus sp.) carrying a large, round load of pollen held on the corbicula. [photo credit: Amber Barnes].

have very dense hairs on their hind legs that hold oils. The males of some species of *Euglossa* have specialized structures on their hind tibia called perfume pouches where they collect floral scents that they then use to attract females. Honey bees and stingless bees collect plant resins, oils, and other secretions that they then mix with their saliva and nectar to make propolis.

3.2.2 Behaviours that optimize pollen collection

Feeding habits vary greatly between insect pollinators, and even within genera, and species of each group. Some species exhibit specialization for one host plant species and are considered monolectic. Others forage on multiple species within one plant genus, a feeding habit that is termed oligolectic. A broader repertoire for multiple species in similar plant genera is called mesolectic. Some have a much broader repertoire and visit multiple species within multiple genera families and are considered polylectic to various degrees [5]. Honey bees, which are considered broadly polylectic or generalist feeders, forage on a wide spectrum of food resources within a landscape. One-to-one specialized relationships where there is a single pollinator for a single plant, though interesting and highlight specialized, are rare. Research into pollination networks within ecosystems has shown that overlap and redundancy are much more common. Bipartite pollinator networks are useful tools to understanding how pollinators and plants within an ecosystem interact, building resilience and buffering against change (**Figure 8**) [6–8].

Some pollinators have specific behaviours they use to increase pollen collection. In some cases the coevolution of this mutualism has resulted in plants that are so specialized they can only be effectively pollinated by certain pollinators. Buzz pollination, or *sonication*, is one such system. In sonication, a pollinator holds on to a flower and vibrates its wings to release pollen. Bumble bees and other large-bodied

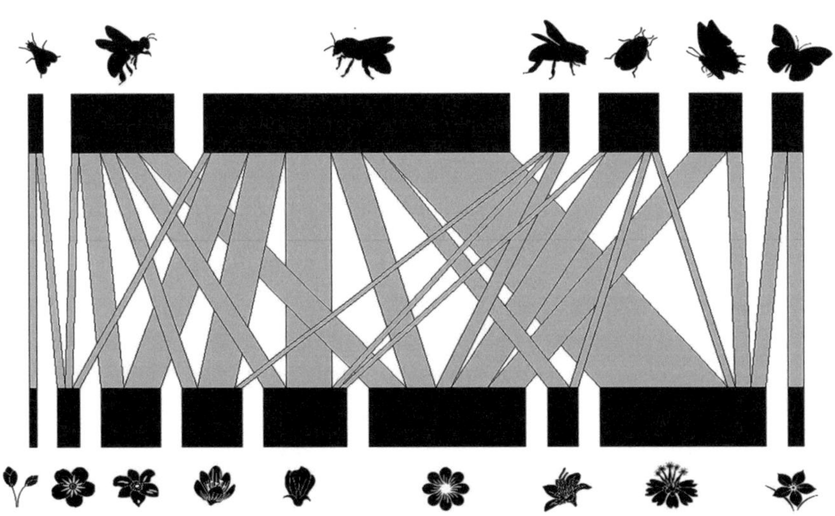

Figure 8.
A representative bipartite network visualizing the linkages and connections in a plant-pollinator system. Linkages between plants and pollinators are indicated with connecting lines, line width represents the relative dominance of the linkage in the system. Bands below pollinators represent relative dominance of each taxon in the system. Similarly, bands above the flowers represent the relative dominance in terms of usage by pollinators in the system. Using these networks allows for keystone species and resilience levels in the pollination systems to be determined, and can also identify plants and pollinators that may be more at risk. [figure credit Victoria Wojcik].

insect pollinators are prime examples. Plants in the family Solanacea (nightshades), which includes tomatoes, eggplants, and potatoes, have adapted to buzz pollination; their anthers are enclosed in a structure and pollen can only be released through vibration.

Bees present the most refined versions of behavioural adaptations to optimize pollen and nectar collection. Flight is metabolically expensive, yet it is the dominant manner in which most insect pollinators find and access their food. Patterns in how pollinators work within and between flowers show both energy optimization through shortened flight patterns and minimizing efforts against gravity, as well as ways to increase the chance of finding nectar. Though they often search for sites of forage, bees are central-place foragers, meaning they tend to forage on food resources nearer to their nest site. For colonial species like honey bees and bumble bees, this means foraging near to their nest. For solitary species this means nesting near abundant food resources. Colonial species with an established nest in a dearth of resources will relocate.

When bees are foraging in a field, they focus the attention of an individual foraging trip on the same species of flower, even if they feed on a diversity of sources. Flight patterns are impacted by resource abundance as an adaptation to optimizing rewards. When flying through a resource-rich floral field they have been documented to make more turns in their flight, stopping more often to feed. In fields with less abundant resources, they fly straight more often, searching for abundant, localized forage. Bees that feed on a wide range of flowers also focus their foraging efforts on the most dominant species in the landscape, a strategy that targets rewards and balances energy expenditure. Optimizing forage to what is in bloom and providing the most nectar and pollen that day minimizes the flight time between nest and flowers and conserves energy. Energy optimization is used for resource seeking and floral selection. Bees tend to use complex floral resources that have vertical structure from the top down, again conserving energy by minimizing the needs to fly against gravity. This is most evident in tropical systems were bees are foraging on flowering trees.

The culmination of behavioural adaptations to optimize foraging is social communication and recruitment foraging exhibited by social bees, such as honey bees, stingless bees, and bumble bees. Bumble bees exhibit rudimentary communication through movement and wing vibrations that appear to signal that the arriving forager has visited profitable food resources. Honey bees and stingless bees have a complicated dance language, commonly called the waggle dance in honey bees. Thorough this dance they communicate the quality and location of food resources. Honey bees provide information on distance and direction, while stingless bees also provide information about the vertical plane in which food is found as many stingless bees forage in forest ecosystems.

4. Globally pollinator diversity

Globally there are estimated to be more than 1,000,000 species of insect pollinators. This includes nearly 800,000 beetles that are not likely to be the main pollinator of most plants, but are the dominant anthophiles.

Speaking to the most effective and functional pollinators, there are currently some 20,000 species of bees known globally. This diversity is represented by seven families: Andrenidae, the mining bees; Apidae, honey bees, carpenter bee, stingless bees, long-horned bees, and orchid bees; Colletidae, the plaster or cellophane bees; Halictidea, the sweat bees; Megachilidea, the leaf-cutter and mason bees; Mellittidae, the oil collecting bees, and Stenotritidae, the fast flying bees. Stenotritidae are only found in Australia; the other families are present across the globe.

Pollinating wasps include members of the subfamily Masarinae, which bare a visual resemblance to vespid wasps. Fig wasps, members of the family Agaonidae, are a diverse group of more than 900 species of tiny wasps that have a unique specialized pollination system with figs (genus *Ficus*). Fig flowers have evolved an internalized structure that is accessible through a small opening. Female fig wasps covered in pollen enter the fig flower to lay their eggs inside, transferring the pollen of the fig that they have hatched from. The eggs of the fig wasp mature and hatch inside of the fig flower. Male and female offspring mate with each other, the males remain in the flower, and the females exit the flower to start the cycle over again.

There are some 150,000 species of moths and 17,000 species of butterflies globally. Many of the relationships that butterflies have with flowers are more specialized, particularly those that include access to nectar in deep nectaries. Both butterflies and moths require host plants for their caterpillar development, in many cases the same plants that provide nectar to adults also serve as larval host plants (**Figures 3–6**).

Though there are roughly 160,000 fly species globally, a narrow spectrum of these species are pollinating anthophiles. The most common group of pollinating flies is the family Syrphidae, with nearly 6000 species. Syrphid flies, also known as flower flies, often resemble bees and mimic these species. Other pollinating flies include midges, such as members of the genus *Forcipomyia*, which pollinate cacao flowers, and mosquitoes, including the snowpol mosquito *Aedes communis* that pollinates the blunt-leaf orchid *Platanthera obtusata*.

Pollinating insects are present in all terrestrial ecosystems, with the exception of Antarctica. Patterns of diversity include both ecosystem types and geographic regions that have higher richness than others. Arid landscapes tend to have higher pollinator species richness, and though this might seem counter intuitive, these ecosystems are characterized by distinct seasonal patterns, often with short precipitation windows, and environmental extremes. This multitude of temporal niches has driven a diversity of pollinator mutualisms. For some species, such as bees, there are distinct biomodal patterns of richness, with far northern and far southern latitudes having less richness than mid-latitudes. In the case of bees, their niche is often filled by flies closer to the poles. This same pattern of declining richness and substitution of flies into bee niches is seen in alpine ecosystems as elevation increases. Meadow and prairie ecosystems are generally considered to be hot-spots of insect pollinator diversity, owning to the richness of flowers and open space. Conversely, temperate forest systems have previously been considered to be lower in pollinator richness, but recent evidence suggests that these systems are in fact abundant with species.

5. Nesting and other habitat needs

Pollinating insects have a diversity of nesting strategies and habitat needs for reproduction. Bees are famous for constructing nests; for social species this including building hives and constructing combs and brood chambers for their young. Honey bees, stingless bees, and bumble bees all produce wax which they use to build nest cells. Solitary bees have a more varied set of nesting strategies. The majority excavate nesting in the soil, creating chambers for brood.

Some species make use of preexisting holes, lining them with various materials. Leafcutter bees cut and fold leaves to make nest cells. Mason bees collect mud to form their nests. Some bees excavate nests in the pith of plant stems, and some, like carpenter bees, are able to chew into wood. Solitary bees provision for their offspring, laying an egg that hatches, eats a pollen store, pupates, and then emerges as an adult.

Butterflies and moths lay eggs on host plants, their young emerge and eat the plant materials, eventually pupating, and emerging as adults. Pollinating flies lay their eggs in leaf litter or other substrates. Beetles have varied nesting lifestyles, including leaf litter and burrowing.

6. Agriculture and food production

Pollinators are responsible for one of every three bites of food we eat [9, 10]. This corresponds to an enormous economic contribution that has been estimated recently to be upwards of 153 billion US dollars annually [11]. Corn, wheat, and rice – the carbohydrate stables of most cultural diets – are wind pollinated, but nearly 80% of the crops that are cultivated across the globe require or benefit from pollinators [10]. Pollinated foods provide the necessary and essential nutrients for the human diet, including foods high in antioxidants, as well as vitamins A, B, and C [12].

The major insect pollinators of agricultural crops are bees and flies, with other insect pollinators generally considered to play minimal roles in crop production. There are many crops, fruits, and spices that are pollinated by moths, wasps, and even beetles, especially in more tropical parts of the world. Butterflies, however, are not known to pollinate any crops.

Honey bees are famous in agriculture because they can be managed and moved between crops to pollinate what is in bloom. As generalist pollinators, honey bees will visit most crops, but they are not universally the ideal pollinator for each crop. Bumble bees are ideal pollinators for tomatoes, eggplants, and peppers as these crops require sonication, or buzz pollination. Squash bees, *Peponapis* spp., pollinate pumpkins, squash, melons, and zucchinis [13]. Orchard and mason bees (*Osmia* spp.) are ideal pollinators of stone fruit (cherries, almonds, etc.), apples, and pears [13]. In the case of pears they outperform honey bees who find pear nectar too low in sugar for their taste. Leafcutter bees (*Megachile* spp.) are good pollinators of alfalfa and other legumes [13]. Vanilla orchids are pollinated by orchid bees in the genera *Euglossa* and *Eulema*.

Flies play key pollination roles in many crops, including staples such as cocoa (midges in the genera *Ceratopogonidae*, *Forcipomyia*, and *Euprojoannisia*), coffee, and tea [13]. Mangos, kolanut, lychee, coconut, nutmeg, acerola, and avocados are pollinated by flies. Flies also pollinate many common field crops such as carrots, onions, leek, parsnip, and dill [13]. Moths have roles in pollinating crops including yucca, neem, papaya, passionfruit, and nutmeg [13]. Wasps are known for pollinating figs, but also play roles in pollinating kenaf, lychee, and cotton [13]. Beetles pollinate pomegranates, parsnip, and nutmeg [13].

7. Additional ecosystem services supported by insect pollinators

Food production and the reproduction of wild plants are considered critical provisional ecosystem services, without which the Earth could not exist in its life-supporting state. It has already been established that insect pollinators are critical to food production and plant reproduction [14]. Additionally, insect pollinators are keystone in provisional, regulating, and cultural ecosystem services.

Hardwood products, fibers, textiles, dyes, scents, plant derived chemicals, and pharmaceutical products are the products of flowering plants that require pollination [15, 16]. Insect pollinators also have a role in maintaining reproduction in plants that fix and cycle nitrogen, such as legumes [17]. Some of the very behaviours and lifestyles of these insects, such as ground nesting, promote soil aeration and nutrient cycling. About 60% of bee species nest in the ground, and this behaviour provides soil disruption [18]. Vegetation that defines ecosystems, and buffers the impacts of severe weather and erosion is again commonly insect-pollinated. The California chaparral ecosystem is one example. Plant species responsible for soil stabilization along hillsides require pollinators for reproduction [19]. A lack of pollinators and the eventual senescence of these species can cause vulnerability to landslides during annual wet periods. Similarly, coastal mangrove communities that provide protection during storm events need pollinators for reproduction [20, 21]. Tropical forest communities, which play significant roles in carbon sequestration, are dominated by pollinator-dependent species, with an estimate of nearly 95% dependence [1].

Cultural inspiration and spiritual enrichment is provided by ecosystems that depend on insect pollinators, as well as the pollinators themselves. Mayan artwork depicts the practice of *meliponiculture*, and this honey was used in sacred rituals and as food. There are multiple references to honey in Eastern and Western religious texts and rituals. The cleansing and rebirthing properties of honey are mentioned heavily in Jewish text and honey is used in celebration of the New Year. In Hinduism honey, *madhu*, is as one of the five sacred elixirs of immortality. In Buddhism honey is given as a gift of honour to monks. Contemporary art and culture also features insect pollinators heavily, be it depictions of bees as mascots for breakfast cereal or our fascination with the metamorphosis of caterpillars into butterflies.

7.1 Managing insect pollination services

7.1.1 Beekeeping

With the significant role that managed honey bees play in global food production their health and wellbeing a key concern for beekeepers, farmers, and governments. Today, honey bees are managed by beekeepers to maintain high quality colonies that meet minimum pollination contract requirements or that produce marketable amounts of quality honey for commercial markets. A conservative estimate of the history of beekeeping dates the practice at 5000–6000 BCE, with images on African pottery and Egyptian artifacts depicting the practice. Modern day beekeeping, the management of honey bees in hives with mobile combs, began in 18th century Europe. In the Americas, the Mayans practiced meliponiculture, keeping stingless bees in the genus *Melipona* for honey that was used in rituals and as a sacred food.

7.1.2 Other managed bees

Other managed bees include multiple species of bumble bees, which can be reared in boxed colonies and used in greenhouse and field pollination. A handful of solitary bees are also reared for commercial use. The alfalfa leafcutting bee, *Megachile rotundata*, is reared in tube nests and used to pollinate alfalfa. Orchard bees such as the blue orchard bee, *Osmia lignaria*, and the red mason bee *Osmia bicornis*, are managed in North America and Europe, respectively, for orchard pollination. The alkali bee, *Nomia melanderi*, is a ground nesting sweat bee that prefers alkali soils. This species also pollinates alfalfa and is commercially managed either in movable binds of soil containing nest cells, or by seeding areas near farms with nests as this species recolonizes it's home range.

7.1.3 Other managed insect pollinators

Many other insect pollinators and other bees are managed by providing nesting and feeding opportunities near to crops of interest. The establishment of floral resource strips near agricultural increases the occurrence of many species of solitary bees, as well as other insect pollinators such as flies, butterflies, and moths. In the case of some fly pollinators, such as those that pollinate cocoa, leaving leaf litter in plantations encourages nesting opportunities and is thought to boost populations and pollination.

8. Threats and challenges

Insect pollinators face survival challenges that stem from multiple, interacting factors. Habitat loss; pests and disease; invasive species; pesticides; and climate change all make survival more difficult for these essential species. These factors work both independently and in synergy to undermine the survival of insect pollinators, and the people and plants that depend on them.

8.1 Habitat decline

The primary threat to insect pollinators is habitat decline, both in terms of the available area, and in the richness of plant species available to feed on. Agricultural intensification, increased urbanization, and the extractive industry disturbs and removes habitat resources. Habitat loss reduces feeding opportunities by both reducing the amount and variety of food sources, and also reduces opportunities for nesting. When habitat loss increases fragmentation reproductive consequences can also occur, include narrowing genetics and reproductive isolation for both pollinators and plants [22].

8.2 Pests and disease

Pests and disease challenge the survival of pollinators, and have increased in their impacts as international travel, commerce, and industrialized agriculture provide increased opportunities for spread. So much of our agricultural productivity is dependent on the honey bee (*Apis mellifera*) that it is no wonder that our attention is drawn to their plight. Honey bees and other bees are subjected to pests and parasites, such as *Varroa destructor*, an external mite that weakens bees by feeding off of their hemolymph. Common diseases include *Nosema ceranae*, a fungal infection that impacts digestion and the absorption of nutrients and foul brood (American foul brood: *Paenibacillus larvae* and European foul brood: *Melissococcus plutonius*), which infect larvae causes mortality. A slate of viruses also infects bees, including deformed wing virus, which causes body malformations; and sacbrood virus, which causes mortality before pupation. These bee diseases have been identified from honey bees, but are known to spread to wild populations of bumble bees and other species, presenting a conservation challenge and further stressor in these unmanaged species.

8.3 Pesticides

Pesticides and other chemical pollutants threaten insect pollinators, in some cases causing direct morality, and in others causing significant sublethal impacts that reduce pollinator function, cognition, and reproduction. The vast majority of targeted pests are insects, and correspondingly beneficial insects within these systems suffer. Most pollinator poisoning occurs when pollinator toxic pesticides are applied to crops during the blooming period. Poisoning of pollinators can also result from drift onto adjoining crops or plants, the contamination of drinking water, and the uptake of systemic pesticides that move through the soil by non-target plant species.

8.4 Invasive species

Invasive species, both plants and other insect pollinators, can alter local ecosystems, in some cases causing significant enough change that pollination webs are modified and individual mutualisms are impacted. Invasive plant species with aggressive, generalist growth patterns can become the dominate species in a landscape. This transition in landscape composition can mean rare resources of more specialized pollinators dwindle. The basic foraging biology of pollinators can also promote the establishment and spread of invasive plants. As generalist pollinators preferentially visit the most dominant species in a landscape, this can provide a feedback loop that exacerbates the issue. Non-native plant species that have not coevolved with the native pollinator fauna may present other challenges such as a phenological mismatch, floral morphology that is difficult to access, and variability in both pollen protein and amino acids and nectar sugar concentration.

Dietary variability can impact growth and reproduction in species that have coevolved with certain food profiles. Invasive arthropod pollinators have also been noted to cause stress on plant-pollinator systems, including competitive interactions with native species that could result in extirpation or extinction, or the transmission of new pathogens and diseases. There are multiple examples of non-native bees that have established outside of their range; they are commonly tube nesters that have been accidentally introduced. One such example is *Megachile sculpturalis*, commonly known as the giant resin bee, this species is aggressive and has been shown to evict other Megachile species from their nests, and even has been aggressive towards carpenter bees (*Xylocopa* spp.). Purposeful introductions for agriculture have occurred as well, including two Asian species, *Osmia cornifrons* and *Pithitis smaragula*, which were introduced to the United States.

8.5 Climate change

Climate change is a looming threat, impacting both insect pollinators directly, and changing the floral ecosystems these species depend on. Changes in richness and diversity; range changes and restrictions; changes in flight periods; as well as asynchrony between coevolved pollinators and plants are the primary concerns as the climate changes. As patterns of precipitation and temperatures change the

DOI: http://dx.doi.org/10.5772/intechopen.97153

range of conditions that define an appropriate niche for insect pollinators and their plants can change.

Overall patterns of pollinator richness have been predicted to change, which includes both increase and decreases in regional richness. For example, predictive modelling of butterfly richness in Canada in a climate change scenario showed decreases at the most northern and southern latitudes, but increase in richness at mid-latitudes [23]. In some cases this will be a range contraction, and in some more severe cases extirpations and extinctions as species are pushed to the extremes for their biology and phenotypic plasticity. Historical museum records of bumble bee occurrence compared to current occurrence data showed a dramatic range restriction in United States [24].

The cues that trigger plant phenology, such as bloom and bud drop, are largely dependent on photoperiod, which remains constant as temperatures change. The maturation and development of insects is largely driven by temperature, with most species requiring a specific number of degree days to pupate and emerge as adults. The development of asynchrony between bloom and the emergence of insect pollinators is another threat; recent studies from Europe indicate that on average there been a two-day dissociation between plants and their pollinators in the past 30 years [25], with some pairings showing as much as 10 days, as is the case between blackcurrant and *Osmia rufa*. Ten days may appear extreme, but a 2011 study of pollinator phenology in the United States found that on average bees are emerging 10 days earlier in the calendar year when compare to historical records from a century ago [26].

9. Conservation actions

Pollinator populations are changing in response to a changing world. The previously mentioned list of threats and challenges has resulted in many insect pollinators being in decline, with many species listed as threatened or endangered. Extinctions in pollinator species have also occurred. When pollinators decline, the plants that they depend on, the productivity of ecosystems, and the services they provide parallel this trend [27]. Conservation actions that aim to support arthropod pollinators include policy frameworks that protect natural areas, moving towards sustainable agricultural systems that include increasing non-crop forage, active programs to protect and boost populations of listed species, policy that works to reduce stressors such as pesticides, and climate positive actions. Key efforts seek to raise awareness of the essential roles that the small, often overlooked and misunderstood, species play in supporting our daily lives.

Author details

Victoria Wojcik
Pollinator Partnership Canada, Toronto, Canada

*Address all correspondence to: vw@pollinator.org

References

[1] Ollerton J, Winfree R, Tarrant S. How many flowering plants are pollinated by animals? Oikos, 2011; 120: 321-326. DOI: 10.1111/j.1600-0706.2010.18644.x

[2] Cai C, Escalona HE, Li L, Yin Z, Huang D, Engel MS. Beetle Pollination of Cycads in the Mesozoic. Current Biology, 2018; 28: 2806-2812. DOI:10.1016/j.cub.2018.06.036

[3] Peñalver E, Labandeira CC, Barrón E, Delclòs X, Nel P, Nel A, Tafforeau P, Soriano C. Thrips pollination of Mesozoic gymnosperms. PNAS, 2012; 109 (22): 8623-8628. DOI: 10.1073/pnas.1120499109

[4] Michener CD, Grimaldi DA. The oldest fossil bee: Apoid history, evolutionary stasis, and antiquity of social behavior. PNAS USA. 1988; 85(17): 6424-6426.DOI: 10.1073/pnas.85.17.6424

[5] Cane JH, Sipes, S. Floral specialization by bees: analytical methodologies and a revised lexicon for oligolecty. In: Waser N, Ollerton, J, editors. Plant-Pollinator Interactions: From Specialization to Generalization. Univ. Chicago Press. 2006. p. 99-122.

[6] Burkle, LA, Irwin, R. The importance of interannual variation and bottom–up nitrogen enrichment for plant–pollinator networks. Oikos, 2009; 118-12:1816-1829 DOI:10.1111/j.1600-0706.2009.17740.x

[7] Burkle, LA,. Delphia, CM, O'Neill, KM. Redundancy in wildflower strip species helps support spatiotemporal variation in wild bee communities on diversified farms. Basic and Applied Ecology, 2020; 44 (1-13) DOI: 202010.1016/j.baae.2020.02.005

[8] Fantinato, E, Del Vecchio, Gaetan, SC, Buffa, G. The resilience of pollination interactions: importance of temporal phases. Journal of Plant Ecology, 2019; 12 (1)157-162, DOI:10.1093/jpe/rty005

[9] Aizen, M. A., Garibaldi, L. A., S. A. Cunningham, and A. M. Klein. 2009. How much does agriculture depend on pollinators? Lessons from long-term trends in crop production. Ann Bot. 103: 1579-1588.

[10] Klein, A. M., B. E. Vaissière, J. H. Cane, I. Steffan-Dewenter, S. A. Cunningham, C. Kremen, and T. Tscharntke. 2007. Importance of pollinators in changing landscapes for world crops. Proc. Biol. Sci. 274: 303-313.

[11] Gallai, N., Salles, J-M., Settele, J., and B. E. Vaissière. 2009. Economic valuation of the vulnerability of world agriculture confronted with pollinatordecline. Ecol Econ. 68: 810-821.

[12] Wojcik, V, Adams, L. Your Health Depends on Pollinators. 2012. Pollinator Partnership.www.pollinator.org/pollinator.org/assets/generalFiles/YourHealthDependsPollinators.brochure.ver7.pdf

[13] McGregor, S.E. Insect Pollination Of Cultivated Crop Plants, originally published in 1976. United States Department of Agriculture [internet].2009. https://www.ars.usda.gov/ARSUserFiles/20220500/OnlinePollinationHandbook.pdf [Accessed: 2021-03-09]

[14] Kremen, C., N. M. Williams, M. A. Aizen, B. Gemmill-Herren, G. LeBuhn, R. Minckley, L. Packer, S. G. Potts, T. Roulston, I. Steffan-Dewenter, et al. 2007. Pollination and other ecosystem services produced by mobile organisms: a conceptual framework for the effects of land-use change. Ecol.Lett. 10: 299-314.

[15] Daily, G.C., editor. 1997. Nature's services: societal dependence on natural ecosystems. Washington, D.C.: Island Press.

[16] Costanza, R., R. d'Arge, R. de Groot, *S. Faber*, M. Grasso, B. Hannon, K. Limburg, S. Naeem, R.V. O'Neill, J. Paruelo, R.G. Raskin, P. Sutton, and M. van den Belt. 1997. The value of the world's ecosystem services and natural capital. Nature 387: 253—259.

[17] Free JB. Insect pollination of crops, 2nd edition. Academic Press. 1993.648 pp.

[18] Michener, C.D. Bees of the World. The John Hopkins University Press, Baltimore. 2000. pp 913.

[19] Kremen, C., R.L. Bugg, N. Nicola, S.A. Smith, R.W. Thorp, and N.M. Williams. 2002. Native bees, native plants, and crop pollination in California. Fremontia 30: 41—49.

[20] Tomlinson PB. The Botany of Mangroves. Cambridge, UK: Cambridge University Press; 1995.

[21] Guannel G, Arkema K, Ruggiero P, Verutes G (2016) The Power of Three: Coral Reefs, Seagrasses and Mangroves Protect Coastal Regions and Increase Their Resilience. PLoS ONE 11(7): e0158094. doi:10.1371/journal.pone.0158094

[22] Aguilar, R., L. Ashworth, L. Galetto, and M.A. Aizen. 2006. Plant reproductive susceptibility to habitat fragmentation: review and synthesis through a meta-analysis. Ecology Letters 9:968-980.

[23] Kharouba, H, Algar, A, Kerr, J. Historically Calibrated Predictions of Butterfly Species' Range Shift Using Global Change as a Pseudo-Experiment. Ecology, 2009; 90(8), 2213-2222.

[24] Kerr JT, Pindar A, Galpern P, Packer L, Potts SG, Roberts SM, Rasmont P, Schweiger O, Colla SR, Richardson LL, Wagner DL, Gall LF, Sikes DS, Pantoja A. CLIMATE CHANGE. Climate change impacts on bumblebees converge across continents. Science. 2015;349(6244):177-80. DOI: 10.1126/science.aaa7031.

[25] Duchenne, F, Thébault, E, Michez, DM, Elias, M, Drake, M, Persson, M, Rousseau-Piot, JS, Pollet, M, Vanormelingen, P, Fontain, C. Phenological shifts alter the seasonal structure of pollinator assemblages in Europe. Nat Ecol Evol 2020; 4, 115-121. DOI:10.1038/s41559-019-1062-4

[26] Bartomeus, I, Ascher, JS, Wagner, D, Danforth, BN, Colla, S, Kornbluth, S, Winfree, R. Climate-associated phenological advances in bee pollinators and bee-pollinated plants PNAS. 2011;108 (51) 20645-20649; DOI: 10.1073/pnas.1115559108

[27] Biesmeijer, JC, Roberts, SPM, Reemer, M, Ohlemuller, R, Edwards, M, Peeters, T, Schaffers, AP, Potts, SG, Kleukers, R, Thomas, CD, Settele, J, Kunin, WE. Parallel declines in pollinators and insect-pollinated plants in Britain and the Netherlands. Science 2006;313:351-354.

Chapter 3

Seasonal Dynamics on Spider Population in Pathiramanal Island, Kerala, India: A Case Study

Jobi J. Malamel

Abstract

Impact of temperature, rainfall, and humidity varied across different seasons, and the spiders responded differently in each season. Spider community reaches its peak in growing season (October to January). The growing season is recorded as the period with average temperature, rainfall, and relative humidity and which is found to be more suitable for spider population to increase, because highest proportion of spiders is trapped during this season. Ecological factors diminished the spider fauna from February to May (dry season) with high temperature and then gradually decreased through June to September (rainy season) because of heavy rainfall. Correlation analysis of variables with species richness and number of individuals is tested to check the statistical significance between them. Season-wise dendrogram is plotted to show the similarity between the seasons. For the estimation of spider diversity in three different seasons, indices such as Fisher alpha diversity index, Shannon diversity index and Simpson's diversity index are evaluated.

Keywords: abiotic factors, diversity, ecology, ectotherms, seasonality

1. Introduction

The driving factors in the variation of species diversity and composition are a major topic in community ecology. Seasonal and daily changes in environmental conditions have a vital role to alter the performance of animals [1]. The environmental conditions have a significant effect on the distribution of animals, and a seasonal variation is clearly visible in the abundance of the invertebrates, especially on ectothermic animals [2, 3]. Ectotherms are strongly influenced by the various biotic and abiotic factors of the environment because the environmental changes affect their distribution and physiology [4]. Though spiders depend on the architectural structure of the habitat in which they live [5], environmental variables such as biotic and abiotic factors cause a resounding consequence on the population dynamics of the spiders. Spiders respond to both the biotic and abiotic factors to facilitate their maximum fitness [6]. Even though spiders are abundant almost in all habitats, they usually opt for a habitat with suitable environmental parameters in order to smooth their growth, reproduction, and survival [6]. The environmental parameters such as temperature, relative humidity and rainfall have influence on the construction of web, survival as well as on selecting the habitat for spiders [7]. All these indicate that the existence of spiders in a

IntechOpen

particular habitat is greatly mediated by different environmental variables on the ecosystem.

A healthy ecosystem and its smooth functioning is an indicator of the potentiality of the biodiversity of that particular ecosystem [8]. Pathiramanal island is a beautiful and tiny Island with an area of approximately 1 km^2 and seems to be a healthy ecosystem with a tower of biodiversity of both plants and animals located in the middle of Vembanad estuary (**Figure 1**). Though small in size, Pathiramanal Island is blessed with rich flora and fauna owing to the presence of wide forest cover and thick vegetation (**Figure 2**). The small size of the Island is the reason to consider it for this case study as small areas give accurate data compared to the larger ones.

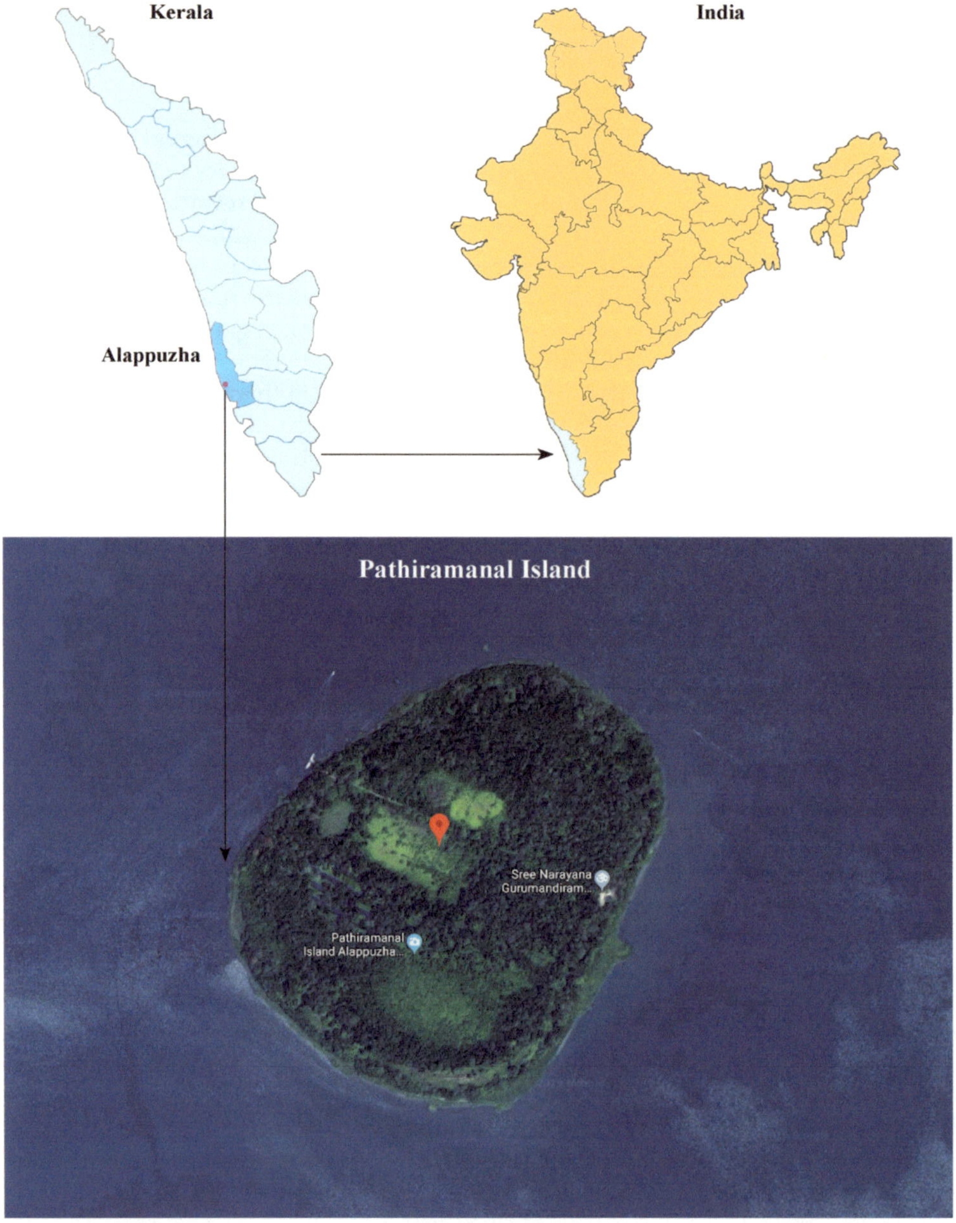

Figure 1.
Map of the study area.

Figure 2.
(A) Pathiramanal Island; (B–G) collection localities of the study area.

Seasonality is an important factor to consider regarding the species distributions in an ecosystem for studies that discuss diversity patterns [9]. In the tropics, seasonal fluctuations have made pronounced variations among the diversity and density of different organisms [10]. In the case of arthropod populations, they are characterised either by changes in species richness and abundance with season or changes not linked with any season [9]. Spider populations have displayed both types of patterns [11], but most of the studies in tropical habitats have recorded a different tendency for each parameter regarding the season for their abundance, composition and community structure [12]. Despite its importance, the seasonal variation of spider communities in studies of tropical diversity has received little attention [13, 14].

So this study is not to deal with the precise amount and composition rather to expose how the diversity and composition of spiders in Pathiramanal Island are influenced by different abiotic factors such as temperature, rainfall and relative humidity. This chapter mainly focuses to check the variation in density and diversity of Araneae in accordance with the change in the selected environmental parameters. The study is also carried out to show the correlation of population dynamics of spiders in three different seasons in the Pathiramanal Island of Kerala, India.

2. Materials and methods

Pathiramanal Island is a small tropical Island with an area of approximately 1 km^2 lies between the latitudes 9°37′07.11″ N and longitudes 76°23′04.95″ E. It is located in the backwaters of Alappuzha District and many rare varieties of migratory birds from different parts of the country come here to nest, adding to the scenic beauty of its location on Lake Vembanad. With respect to its geographical, climatic and ecological features, the Island harbours a rich amount of arachnids of which spiders have a huge share. The temperature ranges from 28.6 to 33.5°C, with an annual mean of 31.0°C. The study was conducted from October 2014 to September 2016 and in the present study, 4 hours of sampling involved active searching for spiders employing a combination of five collection methods such as aerial hand collection, ground hand collection, litter sampling, sweep netting and vegetation beating. All the collected specimens during the survey transferred to the fixative (70% alcohol) for preservation. The sex and developmental stage of all trapped individuals are determined in the laboratory. Species-level identification is mainly recognised by looking at the genitalic features of the spiders. The palp and epigyne are dissected and cleared in 10% KOH for identifying the species. The identification and classification are done also based on the various body parts. A detailed taxonomic study is carried out using the data provided by the World Spider Catalog [15].

The investigation is done to determine how the abiotic factors influence the population dynamics of spiders in Pathiramanal Island. In order to study the variation in population density of spiders with different abiotic factors, a random collection was made in the habitat using the sampling techniques such as hand picking, beating and pitfall traps, and the number of spiders present in the Island is recorded. The collection was done once in a month from morning 8.00 am to 12.30 pm noon throughout the year. Various climatic parameters viz. temperature, relative humidity and rainfall are recorded during the period of collection.

The entire year is divided into growing season (GS), dry season (DS) and rainy season (RS). The data are subjected to statistical analysis and correlation between the spider population and abiotic factors in every three seasons are derived. Pearson's correlation coefficient (Rp) is calculated to study the correlation between different abiotic factors, viz. mean temperature (°C), relative humidity (%) and rainfall (mms) on the species richness and number of individuals using software PAST version 3.19. For the estimation of spider diversity in three different seasons, indices such as Fisher alpha diversity index, Shannon diversity index and Simpson's diversity index are evaluated.

2.1 Data analysis

Statistical relevance of the collected data was supported by calculation of the following diversity indices:

Shannon-Weiner diversity index (H) is calculated using the formula,

$$\mathbf{H} = -\Sigma\left[(\mathbf{pi}) \times \mathbf{ln}(\mathbf{pi})\right] \tag{1}$$

where Σ = summation and pi = proportion of total sample represented by species i.

Simpson's diversity Index (D' = 1 – D) is calculated using the formula:

$$D = 1 - \sum n(n-1) / N(N-1) \tag{2}$$

$$D' = 1 - D = 1 - \frac{\sum n(n-1)}{N(N-1)} \tag{3}$$

where D = Simpson's index, n = the total number of organisms of a particular species, and N = the total number of organisms of all species.

Evenness in species distribution is calculated using Simpson's formula,

$$\mathbf{E} = \mathbf{D} / \mathbf{Dmax}, \tag{4}$$

where

$$D = 1 / \Sigma Pi\,2 \tag{5}$$

Chao1: an estimate of total species richness is calculated using the formula:

$$\text{Chao1} = S + F1(F1-1)/(2(F2+1)), \tag{6}$$

where F1 = number of singleton species and F2 = number of doubleton species.

3. Results

The species composition of the spiders inhabiting Pathiramanal Island varied seasonally and each season responded to the environmental variables differently. During GS, the number of individual spiders collected from the Island was 2405, and the observed species richness, Shannon diversity and Simpson's (1-D) diversity were 129, 4.041 and 0.971, respectively. The number of individuals for DS was 1367, and the observed species richness, Shannon diversity and Simpson's (1-D) diversity were 98, 3.81 and 0.963, respectively. The number of individuals for RS was 965, and the observed species richness, Shannon diversity and Simpson's (1-D) diversity were 77, 3.64 and 0.961, respectively. It is evident from the unstandardized reference sample that GS appears to have higher observed species richness, Shannon diversity and Simpson's (1-D) diversity than others. The abundance of spiders also showed differences in the three seasons, GS showed highest abundance at 521.66 ± 311.73 (mean ± SD, n = 3), followed by DS (213.33 ± 141.50) and RS reported the lowest

(143 ± 78.63) abundance. The spider assemblages during GS showed the maximum Shannon diversity and Simpson's (1-D) diversity, RS appears to have lower Shannon diversity than GS and DS, while DS has lower diversity than GS. The overall seasonal trends in abundance of spider population in Pathiramanal Island shows a sharp decline in the rainy season and dry season, but peak in the growing season. Seasonal dynamics of the spider population in different seasons during the study revealed that the spider population attained its peak in the growing season (October to January). After the growing season, the density of spiders gradually decreased due to adverse climate condition of high temperature (**Figure 3**), while in the rainy season, the lowest numbers of spider species are observed (**Figure 4**). The study revealed that species richness varies in accordance with the relative humidity (**Figure 5**). GS is recorded as the period with average temperature, rainfall and relative humidity and which is found to be more suitable for spider population to increase because the

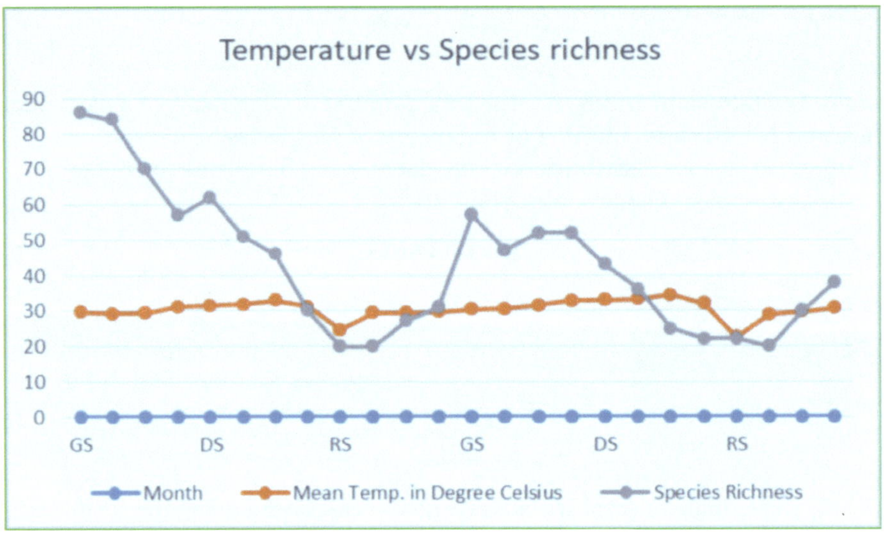

Figure 3.
Effect of temperature on species richness.

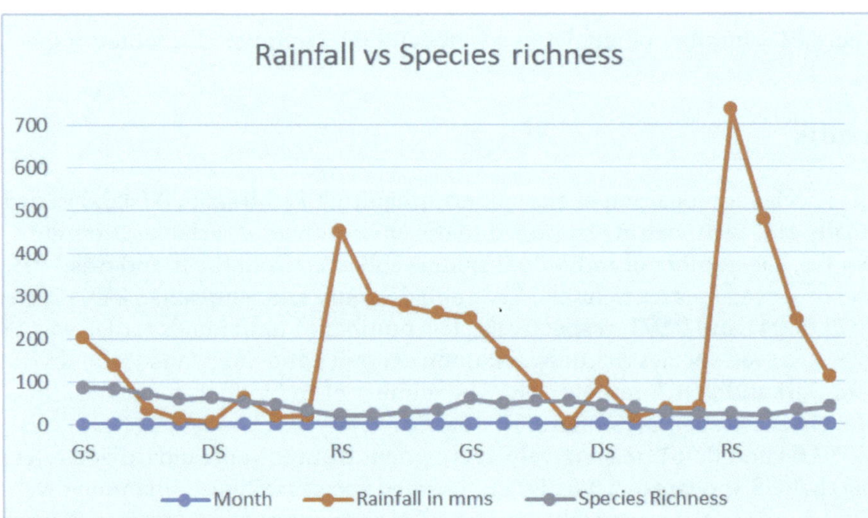

Figure 4.
Effect of rainfall on species richness.

highest proportion of spiders (50.77%) is trapped during this season. Ecological factors diminished the spider population (28.85%) from February to May (dry season) with high temperature and then gradually decreased (20.37%) through

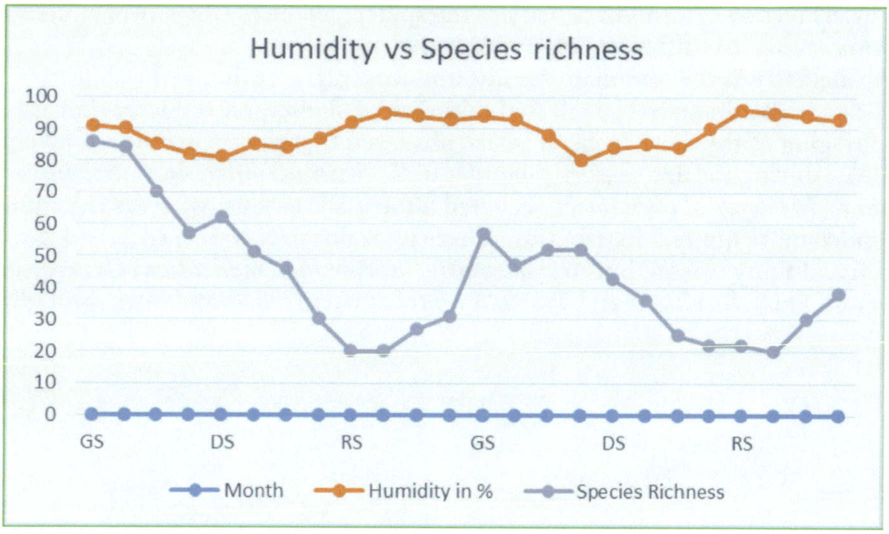

Figure 5.
Effect of relative humidity on species richness.

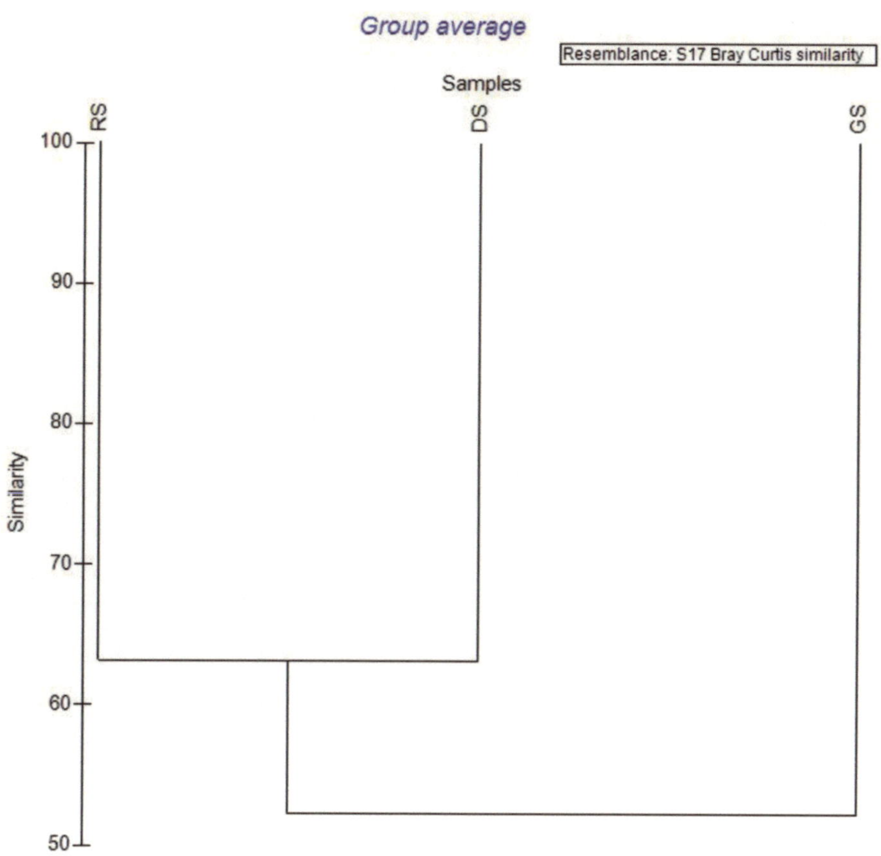

Figure 6.
Season-wise dendrogram.

June to September (rainy season) because of heavy rainfall. Although the correlation between the ecological parameters and species richness is visible from the three graphical representations, statistically no significance is shown. Season-wise dendrogram shows the spider diversity of rainy season (RS) and dry season (DS) are closely related in terms of density of the spiders, while the density of spiders in growing season was different (**Figure 6**).

Araneidae was the dominant family, and this might be due to the collecting localities of Pathiramanal Island that take care of shelter, reproductive behaviour and foraging of these orb webs. It is also observed that the occurrence of juveniles was least during the dry season. Juveniles of *Thelcticopis virescens*, *Carrhotus viduus* and *Heteropoda venatoria* occurred almost all seasons, whereas the adults had much more limited distribution. There were no species with strictly dry season and rainy season, but *Argiope pulchella*, *Anepsion maritatum*, *Gasteracantha geminata*, *Phintella vittata* and *Tylorida ventralis* occurred abundantly throughout

Figure 7.
(A) Argiope pulchella*; (B)* Anepsion maritatum*; (C)* Gasteracantha geminata*; (D)* Phintella vittata*; (E)* Tylorida ventralis*; and (F)* Loxosceles rufescens.

Environmental variables	Species richness		Number of individuals	
	Pearson's correlation coefficient (Rp)	p-value	Pearson's correlation coefficient (Rp)	p-value
Mean temperature (in °C)	0.192	0.363	0.137	0.520
Rainfall (in mm)	-0.420	0.041	-0.387	0.061
Humidity (in %)	-0.380	0.066	-0.300	0.153

Table 1.
Correlation analysis of the variables.

the year (**Figure 7A–E**). The presence of the poisonous spider *Loxosceles rufescens* (**Figure 7F**) in the Island reminds that the tourists should be aware of the spider because of their ability to cause skin necrosis or loxoscelism with their necrotizing venom. It was interesting to note that lycosid spiders made their dominance mainly during the rainy season. Kumari et al. observed that rainfall negatively affected the spider population.

Correlation analysis of the variables such as mean temperature, rainfall and humidity with species richness and a number of individuals revealed statistically no significant relationship between them except those between rainfall and species richness which has been found to have a weak negative correlation (-0.42, $p < 0.05$). However, as indicated by the Pearson's correlation coefficient (Rp) (**Table 1**), species richness and number of individuals tend to be positively correlated with mean temperature, whereas species richness and a number of individuals are negatively correlated with humidity and rainfall.

4. Discussion

The results suggest that abiotic factors and the environment play an important role in influencing the seasonality in abundance, richness and variation of spider species in Pathiramanal Island. The impact of these factors further varied among different habitats within the Island. As a general trend, the species richness was high during the growing season and decreased during the dry season, being minimum during the wet season. The abiotic factors such as temperature, humidity and rainfall influenced the ambient habitat conditions such abundance and humidity of litter, prey arthropod composition and the composition of plant species, thereby tending to reflect on the abundance, diversity and seasonality of spiders.

Habitat preference is very crucial for spiders since it plays a profound impact on the growth, reproduction and fitness of spiders [16]. Habitat utilisation of spiders is always associated with environmental factors such as temperature, wind, rain and humidity [7, 17]. Spiders are plentiful in their distribution where their habitat conditions are favourable [18, 19]. It is observed that spiders choose their habitats in view of the prey availability [20–22], and when the prey availability is less, the spiders move to other habitats with sufficient preys [23]. So, site selection is very important for spiders, and it is greatly influenced by the biotic and abiotic environmental parameters [24].

Results clearly demonstrate that an optimum temperature and relative humidity favoured spider population to increase through (October–January) on the one hand; on the other hand, increase in temperature, humidity and rainfall suppressed spider population in the other months. This study revealed the relative importance of diverse habitat types on diversity and composition of spider assemblages in the Pathiramanal

Island. The habitat heterogeneity hypothesis states that the more complex the habitat, the higher the species diversity and structure. Habitat covariates viz. humidity, temperature and rainfall were found to be important predictors for spider assemblages, and the effect of these variables varied across different seasons [25].

Being ectotherms, the temperature is found to have a key role in the habitat choice of spiders, as they are considered to be constrained by their thermal environment [26]. A recent study [27] reports that the increase in overall temperature due to global warming has resounding consequences on spider populations. The scientists also observed that spiders are unable to withstand the high temperature as well as they need a favourable temperature to cope up with the environment. Li and Jackson [28] learned that spiders adapted to warmer climates can withstand higher temperatures and reproduce at a faster pace, while spiders living in cooler climates develop sooner in response to cooler climates and more slowly at higher temperatures. They, therefore, suggested spiders have evolved to adapt to their natural environments and temperature can act as a regulator in prohibiting optimal foraging within a habitat. A wider temperature range would support greater diversity and abundance [29].

Humidity is another factor regulating the population dynamics of the spiders in an environment. Humidity crucially influences the moulting of the spiders because several moults happen in the lives of the spiders. It is studied that spiders prefer an optimum humidity prior to their moulting, because too much humidity creates fungus and too little will cause moult to go wrong [1]. Both of these conditions can result in the death of the spider.

Riechert and Tracy [16] discovered that extreme thermal stress in some areas prevent the spider from being active and resulted in a lack of prey availability. Therefore, the immature species are not fit enough to withstand the temperature in the dry season, while the adult specimens are better enough to withstand the dry season rather than small spiders.

Lycosid spiders are known as active thermoregulators since they are able to increase their body temperature in accordance with the changes in the environment [30]. Tracey [8] reported that lycosid spiders were very abundant in the high rainfall sites since they are well adapted to this type of environmental disturbance. Rainfall can alter the foraging activity of animals by precluding the visual or tactile perception of prey or predators nearby, or by hindering the mobility of small animals [31].

The season-wise dendrogram showed that species composition of spider assemblage during the DR and RS were relatively similar compared to that of the GS, both showing declining species richness, owing to non-ambient habitat conditions of the respective seasons (high temperature and heavy rainfall, respectively).

5. Conclusion

The present study found distinct compositions of spider species in the three defined seasons sampled, with seasonal distribution throughout the year. The effect of climatic variables, mainly relative humidity, temperature and rainfall, but many other factors need to be investigated, including the diversity of hunting strategies and habitat selection practiced by the animals, characteristics of the vegetation, prey availability and natural enemies. It is concluded that in Pathiramanal Island, growing season is very favourable to spider population due to its variable climatic factors and prey availability. This is the first long-term study investigating spider diversity and its relationship with seasonal variation and habitat distribution in this Island. With this work, it is intended to provide evidence of the possible usefulness of seasonal dynamics with spiders in the Island, testing it as a species richness predictor. It is to consider the effects of environmental factors to test the use of this kind of approach as a tool for conservation efforts.

This is the first intensive sampling implemented to investigate the seasonal response of spiders through different seasons in this small Island. The analysis of spider assemblages together with seasonal variation is a useful approach for understanding mechanisms shaping spider diversity in Island ecosystems. The results demonstrated that seasonal patterns of spider assemblages can differ noticeably revealing a complex spatiotemporal dynamic in this community. Undoubtedly, other habitat elements, not measured in this study, have influenced the seasonal responses of the spider community in the Island.

The results of the present study showed how the population density of spider varies with seasonal changes which, in future, can be used as a base for selecting spider as a bioindicator in a particular ecosystem. It can be used as baseline data to collect the spiders from this site in accordance with the change in the environment. It is also important to note that spider fauna is ubiquitous in nature, and their diversity cannot be explained by quantifying the seasonality of the environment. It does depend on many other factors or a combination of factors such as altitudinal variation, habitat structure, competition, predation, habitat type, environmental stability and productivity. Looking into these factors would surely bring in more interesting results, which can be relevant for stability and management of spider diversity of this region. It is recommended to have extensive surveys of spiders in this unique ecosystem which are important for a better understanding of the seasonal dynamics of such habitats, which is required to support the sustainable management of both the spiders and Island.

Author details

Jobi J. Malamel
Division of Arachnology, Department of Zoology, Sacred Heart College (Autonomous), Thevara, Cochin, Kerala, India

*Address all correspondence to: jomalamelcmi@gmail.com

References

[1] Opell BD, Karinshak SE, Sigler MA. Humidity affects the extensibility of an orb-weaving spider's viscous thread droplets. Journal of Experimental Biology. 2011;**214**:2988-2993

[2] Brown GP, Weatherhead PJ. Thermal ecology and sexual size dimorphism in northern water snakes, *Nerodia sipedon*. Ecological Monographs. 2000;**70**:311-330

[3] Huey RB. Temperature, physiology, and the ecology of reptiles. Biology of the Reptilia. Physiological Ecology. 1982;**12**:25-91

[4] Huey B, Stevenson RD. Integrating thermal physiology and ecology of ectotherms: A discussion of approaches. American Zoologist. 1979;**19**:357-366

[5] Mac Arthur RH, Diamond JM, Karr JR. Density compensation in island faunas. Ecology. 1972;**53**:330-342

[6] Scharf I, Ovadia O. Factors influencing site abandonment and site selection in a sit-and-wait predator: A review of pit-building antlion larvae. Journal of Insect Behavior. 2006;**19**:197-218

[7] Gillespie RG. The mechanism of habitat selection in the long-jawed orb-weaving spider *Tetragnatha elongata* (Araneae, Tetragnathidae). Journal of Arachnology. 1987;**15**:81-90

[8] Tracey BC. Spiders as ecological indicators in the Australian tropics: Family distribution patterns along rainfall and grazing gradients. In: Proceedings of the 17th European Colloquium of Arachnology; 1997; Edinburgh. 1998. pp. 325-330

[9] Kishimoto-Yamada K, Itioka T. How much have we learned about seasonality in tropical insect abundance since Wolda (1988)? Entomological Science. 2015;**18**:407-419

[10] Garcıa E. Modificaciones al sistema de clasificaci'on clim'atica de K¨oppen (para adaptarlo a las condiciones de laRep'ublica Mexicana). UNAM, Ciudad de Mexico; 2004

[11] Torres-Sanchez MP, Gasnier TR. Patterns of abundance, habitat use and body size structure of *Phoneutria reidyi* and *P. fera* (Araneae: Ctenidae) in a Central Amazonian rainforest. The Journal of Arachnology. 2010;**38**:433-440

[12] Martin L, Philpott SM, De la Mora A, Ibarra-Nunez G, Tryban S, Perfecto I. Response of ground spiders to local and landscape factors in a Mexican coffee landscape. Agriculture, Ecosystems and Environment. 2016;**222**:80-92

[13] Campuzano EF, Ibarra-Nunez G, Chame-Vazquez ER, Montano-Moreno H. Understory spider assemblages from a cloud forest in Chiapas, Mexico, and their relationships to environmental variables. Arthropod-Plant Interactions. 2016;**10**:237-248

[14] Viera C, Gonzaga MO. Behaviour and Ecology of Spiders, Contributions from the Neotropical Region. Switzerland: Springer; 2017

[15] World Spider Catalog. World Spider Catalog. Version 21.5. 2020. Natural History Museum Bern. DOI: 10.24436/2. Available from: http://wsc.nmbe.ch [Accessed: 12 May]

[16] Riechert SE, Tracy CR. Thermal balance and prey availability: Bases for a model relating web-site characteristics to spider reproductive success. Ecology. 1975;**56**:265-285

[17] Campuzano EF, Ibarra-Núñez G, Machkour M, Rabet S, Morón-Ríos A, Jiménez ML. Diversity and seasonal variation of ground and understory

spiders from a tropical mountain cloud forest. Insect Sci. 2019;**27**:826-844

[18] McNett JB, Rypstra AL. Habitat selection in a large orb-weaving spider: Vegetation complexity determines site selection and distribution. Ecological Entomology. 2000;**25**:423-432

[19] Orions GH, Wittenberger JF. Spatial and temporal scales in habitat selection. American Naturalist. 1991;**137**:29-49

[20] Harwood JD, Sunderland KD, Symondson WOC. Web location by linyphiid spiders: Prey specific aggregation and foraging strategies. Journal of Animal Ecology. 2003;**72**:745-756

[21] Rypstra AL. Aggregations of *Nephila clavipes* (Araneae, Araneidae) in relation to prey availability. Journal of Arachnology. 1985;**13**:71-78

[22] Thevenard L, Leborgne R, Pasquet A. Web-building management in an orb-weaving spider, *Zygiella notata*: Influence of prey and conspecifics. Comptes Rendus Biologies. 2004;**327**:84-92

[23] Nakata K, Ushimaru A. Difference in web construction behaviour at newly occupied web sites between two *Cyclosa* species. Ethology. 2004;**110**:397-411

[24] Miyashita T. Contrasting patch residence strategy in two species of sit-and-wait foragers under the same environment: A constraint by life history? Ethology. 2005;**111**:159-167

[25] Chapman DS, David MG. Global warming – more than hot air. Journal of Land, Resources and Environmental Law. 2007;**27**(1)

[26] Cobb VA. Effects of temperature on escape behaviour in the cribellate spider, *Oecobius annulipes* (Araneae, Oecobiidae). The Southwestern Naturalist. 1994;**39**:391-394

[27] Kanyama AC, Svahn PW, Sonnek KM. "We want to know where the line is": Comparing current planning for future sea-level rise with three core principles of robust decision support approaches. Journal of Environmental Planning and Management. 2019;**62**(8):1-20

[28] Li D, Jackson RR. How temperature affects development and reproduction in spiders. Journal of Thermal Biology. 1996;**21**:245-274

[29] Petcharad B, Miyashita T, Gale GA, Sotthibandhu S. Spatial patterns and environmental determinants of community composition of web-building spiders in understory across edges between rubber plantations and forests. The Journal of Arachnology. 2016;**44**:182-193

[30] Aspey WP, Lent C, Meeker M. Effect of humidity on desiccation by living and dead wolf spiders (Araneae, Lycosidae). Experientia. 1972;**28**:1249-1250

[31] Queiroz M, Gasnier TR. Strong negative effect of diurnal rainfall on nocturnal activity of a wandering spider in Central Amazonia. Revista de Biología Tropical. 2017;**65**:1152-1160

Chapter 4

Bacterial Disease Control Methods in Shrimp (*Penaeus*, 1798) Farming Sector in Asian Countries

Jeyachandran Sivakamavalli, Kiyun Park, Ihn–Sil Kwak and Vaseeharan Baskaralingam

Abstract

Aquaculture industry produces the enormous amount of sea foods (fish, shrimp, planktons, etc.) with enriched quantity of proteins, essential amino acids, essential fatty acids, and micronutrients and also possesses the medicinal values. This production industry is very important to meet out the need of the global population. Recently, different culture practices for aquatic culturing organisms were developed in practices, where the risk of infection and diseases outbreak also increased which leads to the production loss to the aquatic sector. Several conventional methods are used to prevent the diseases probiotics, antibiotics, plants, immmunostimulants, proteins, immune proteins enhancement, nanoparticles, etc. At the same time, these treatment techniques also have merits and demerits to execute into the practical platform. For instance, chemical or antibiotics treatment into the culture system leads to the some adverse effects in culturing organisms, environment, and also consumer. In this chapter, various diseases caused by the bacterial strains and its control strategies in the shrimp farming industry to enhance the aquaculture are discussed.

Keywords: aquaculture, pathogens, plants, disease management, immunostimulant

1. Introduction

Shrimp farming plays the major role in aquaculture industry globally; due to its proteinaceous nature increased export viability and high profit yield the enhanced economy to the country. Penaeid (Rafinesque, 1815) shrimp aquaculture is one of the major industries which have rapidly grown during the past three decades in tropical and subtropical areas of the world (FAO 2019). Global production of shrimp increased from 1,564,563 metric tonnes in 2017 to 2,002,449 metric tonnes in 2019 (FAO 2019). The black tiger shrimp, *Penaeus monodon* (Fabricius, 1798), Indian white shrimp, *Fenneropenaeus indicus* (H. Milne Edwards, 1837) and Pacific white shrimp *Litopenaeus vannamei* (Boone, 1931) are important commercial species of the Penaeidae family. *F. indicus* supports commercial fisheries in both marine and estuarine environments on the east and west coasts of India. India has been a major supplier of shrimp to Japan, Europe, and USA. In India, Indian white shrimp *F. indicus*, Pacific white shrimp *L. vannamei,* black tiger shrimp *P. monodon,* white shrimp

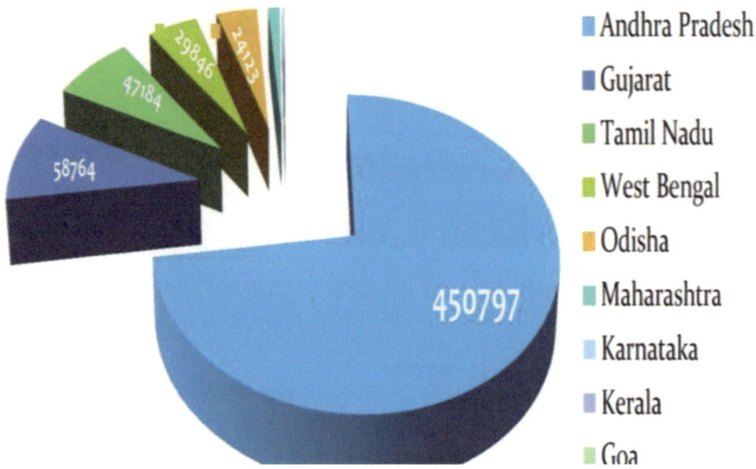

Figure 1.
Aquaculture sector contribution towards economy throughout India.

P. penicillatus (Olivier, 1791), green tiger shrimp *P. semisulcatus* (De Hann, 1844) and banana shrimp *P. merguiensis* (De Man, 1888) are farmed along the coastal areas. Although the production of cultured shrimp has increased, there have been considerable periodic losses due to disease of the farmed shrimp. Global Aquaculture Alliance (GAA) survey with respect to agents responsible for diseases has already revealed that 50% of losses due to diseases were attributed to viruses and about 22% to bacteria [1]. The global loss of shrimp production leads to research against the control of diseases required for the stability of aquaculture industry. In Indian aquaculture industry produces at the Global Outlook in Aquaculture Leadership (GOAL) conference, held in October in Chennai, India, the forecast was for Indian production to drop in 2019. The GOAL prediction has India flat at around 600,000 tonnes in 2019 and 2020, down from as much as (**Figure 1**).

The shrimp immune system, like other invertebrates lacks an adaptive immune system and relies solely on its innate immunity against invading pathogens. Innate immunity is an ancient protective mechanism that appeared early in the evolution of metazoans and is divided into humoral and cellular responses [2], which work in jointly coordination for the detection/elimination of all foreign organisms potentially hazardous for the host [3]. The cellular response mediated by haemocytes in hemolymph involves nodule formation, phagocytosis, encapsulation of pathogens and coagulation [4, 5]. The humoral components include the activation and release of molecules stored within haemocytes, such as anticoagulant proteins, agglutinins, phenoloxidase enzyme, antimicrobial peptides and protease inhibitors [3].

2. Commercially important shrimps in India

2.1 *Fenneropenaeus indicus*

The Indian white shrimp, *F. indicus* formerly known as *P. indicus* is a marine shrimp, prefers mud or sandy-mud bottom and can be found from 2 to 90 m depth. It attains up to 228 mm (nearly 9 inches) in length up to 14–20 g in weight and can tolerate low water quality, high salinities and high temperatures. It is one of the most important Indian commercial species, especially for the inshore fisheries and for rice field culture in Kerala and also captured in the East African coast. The taxonomic position of the Indian white shrimp *F. indicus* is.

Phylum: Arthropoda
Subphylum: Crustacea
Class: Malacostraca
Order: Decapoda
Suborder: Dendrobrachiata
Family: Penaeidae
Genus: *Fenneropenaeus*
Species: *indicus*

2.2 *Litopenaeus vannamei*

Phylum: Arthropoda
Subphylum: Crustacea
Class: Malacostraca
Order: Decapoda
Suborder: Dendrobrachiata
Family: Penaeidae
Genus: *Litopenaeus*
Species: *vannamei*

In penaeid shrimp farming bacterial diseases are commonly associated with natural microbial flora of seawater, which possess enriched organic matter that supports the growth and multiplication of bacteria and other microorganisms. The most common shrimp pathogenic bacteria belong to the genus *Vibrio*. Other Gram-negative bacteria such as *Aeromonas* spp., *Pseudomonas* spp., and *Flavobacterium* spp., are also occasionally implicated in shrimp diseases.

3. Bacterial Septicaemia (*Vibrio* disease)

Acute hepatopancreatic necrosis disease (AHPND) is one of the severe systemic diseases caused by bacteria *Vibrio parahaemolytics* especially in shrimps such *P. monodon* and *P. vannamei*. In this disease the infected animals periopods are red color in nature owing to its chromatophores, sometimes the severe infections of bacterial diseases in shrimps may occur, while gills looks eroded and melanization took place to form the black blisters can be seen on the carapace and abdomen. Apart from *V. parahaemolyticus* some other bacterial pathogens such as *V. alginolyticus, V. anguillarum, V. parahaemolyticus, Vibrio* spp. also involved in this disease pathogenesis.

3.1 Luminescent bacterial disease

This bacteria also causes the dangerous problems in aquaculture farms and heavy losses due to its infections it leads to the economy downfall and production rate loss too. These luminescent bacteria infected shrimps could be look like a fluorescent or luminescent producing nature in darkness. *V. harveyi* are the major pathogens creating this problem in hatcheries. The luminescent bacteria could be isolated using Zobell's Marine Agar, followed by morphological and biochemical characteristics.

3.2 Brown spot disease (Shell disease or rust disease)

Infected animals showed the brown and black erosions on the surface of the body and whole body appendages, this could be caused through *Vibrio* spp., *Aeromonas* spp., and *Flavobacterium* spp., with chitinolytic activity. Diagnosis could be achieved by simple observations such as gross signs and confirmed by isolation of the bacteria from the site of infection on Zobell's Marine Agar and identification of the pathogen.

3.3 Necrosis of appendages

The tips of walking legs, swimmerets and uropods of affected shrimp undergo necrosis and become brownish and black. The setae, antennae and appendages may be broken and melanised. The epibiotic bacteria such as *Vibrio* spp., *Pseudomonas* spp., *Aeromonas* spp. and *Flavobacterium* spp., produced the gross signs in infected shrimps.

3.3.1 Vibriosis in shrimp larvae

The affected larvae show necrosis of appendages, expanded chromatophores, empty gut, absence of fecal strands and poor feeding. Cumulative mortalities may be very high reaching up to 80% within few days. *V. alginolyticus, V. parahaemolyticus, and V. anguillarum* caused this disease.

3.4 Filamentous bacterial disease

The affected shrimp larvae show fouling of gills, setae, appendages and body surface. Molting of affected shrimps is impaired and may die due to hypoxia. *Filamentous* bacteria, such as *Leucothrix mucor* are the causative agent for this disease. Diagnosis of filamentous bacterial disease could be achieved based on gross signs and symptoms and by microscopically demonstrating filamentous bacterial fouling of body surface and appendages of shrimp larvae.

4. Control measure of shrimp disease

Most common pathogenic bacteria of penaeid shrimp include *Vibrio* sp., *Aeromonas* sp., *Photobacterium* sp., *Citinoclastic* sp., *Leucothrix* sp. and *Thiothrix* sp. The loss of a stable microbial balance through disinfection leads an environment wide open for the proliferation of any opportunistic bacteria [6]. Therefore, disease control strategies ought to be a priority in the aquaculture practices. Several antibiotics are used in the aquaculture practices for treatment and to control diseases. The prolonged application of antimicrobial agents at sublethal concentrations may provoke the adaptation of microorganisms to antimicrobial agents [7].

5. Antibiotics

Antibiotics are potential molecules for the initial treatments, though it has its own demerits such as continues usages of antibiotics in the environment like farms, aquatic systems might be causes the pollution and also leads to the development of multiple drugs resistant strains in the environment [8]. For instance, adequate usage of chloramphenicol in shrimp farming sector in Myanmar, India, Pakistan, and Vietnam, paves the way to abuse of drugs resulting heavy loss in farming sector [9]. Considering the high promising results obtained in the in vitro screening of commercial antibiotics, the post-infection therapy using antibiotics remain the method of choice for many farmers [10]. Use of microbes for beneficial purposes is increasingly recognized as a valuable input for sustainable aquaculture. Nowadays, several environmental-friendly prophylactic and preventive methods like probiotics, immunostimulants, antimicrobial peptides and quorum sensing interference are developed to control aquatic organism diseases. Therefore, novel antimicrobials with increased potency and least residual accumulation in shrimp tissue are required in lieu of conventional antibiotics for the management of bacterial epizootics. To keep the shrimp farming as a sustainable venture, new health management strategies must be used instead of the traditional methods like the abuse of antibiotics and chemotherapeutics.

5.1 Herbs as antibiotics

Herbs act as antibiotic for controlling or reduce the infection of pathogen in aquaculture sector and also increase the survival rate of organisms, during outbreak of disease managements. In *Fenneropenaeus indicus*, the anti-vibrio disease controlled by garlic extract [11]. Hot water extracts of brown seaweeds *Sargassum* sp. act as antibiotic against white spot syndrome virus in shrimp *P. monodon* [12]. *Azadirachta indica* plant extract act as antibiotics for treating *Citrobacter freundii* bacterial infection in *Oreochromis mossambicus*. Castro [13] observed, methanoic extract of Brazil herbs act as disinfectant against fish pathogens such as *Streptococcus agalactiae, Flavobacterium columnare* and *Aeromonas hydrophila.* In *Catla catla* disease resistance developed by fish immersed in extract of three herbs namely *Allium sativum, Azadirachta indica* and *Curcuma longa* [14]. The majority of herbs act as anti-pathogenic agent, acts as antibiotic due to strengthen the immune system of organisms prevent from disease or forming disease resistance variety in aquaculture sector.

6. Vaccination

Vaccination is the practice of administering weakened or dead pathogenic bacteria, in order to confer long lasting protection through immunological memory [15]. Adaptive secondary memory immune response of vertebrates depends on immunoglobulins (Igs), T-cell receptors (TCRS), major histocompatibility complex (MHC) and memory T cells. Memory cells and adaptive immunity differentiates the vertebrate and invertebrates immunity. Hence the several strategies are used to improve the adaptive immune system of invertebrates. Vaccination strategy must be designed with the key considerations of minimizing immunomodulatory stresses and stimulates the host defenses by triggering specific immune responses against infectious diseases.

7. Immunostimulants

Immunostimulants are chemical or natural source compounds that activate the immune system of aquatic animals and make them more resistant to infections by viruses, bacteria, fungi, and parasites. Stimulation of the non–specific immune system can improve the animal's response to challenges from pathogenic bacteria. Immunostimulants used to control vibriosis in shrimp increased the survival rate [16, 17]. The potential of immunostimulants is to reduce the effects of bacterial diseases and to improve larval growth. Nowadays commercial immunostimulants are produced in the aquaculture sector to reduce the microbial diseases, through potential activity, immunostimulating performance are not in satisfied level.

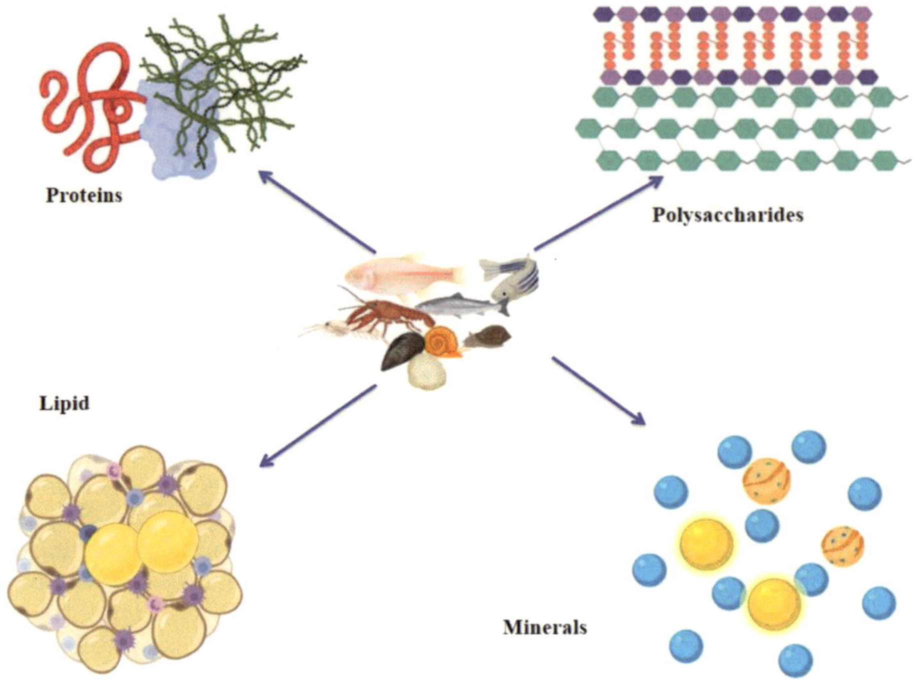

Figure 2.
Marine animal possessing the distinctive level of biomolecules.

Immuno stimulation might be too drastic and harm or even kill the host. Because there is no memory component involved, the response is likely to be short in duration, and hence immunostimulants have to be administered repeatedly. In addition, long term administration of such agents seems to decrease the immune stimulatory effect and does not always promote disease resistance [18]. Several bioactive compounds are isolated through various marine animals' body components (**Figure 2**).

8. Probiotics

During the past two decades, the use of probiotics as an alternative to antibiotics has shown to be promising in aquaculture, particularly in fish and shellfish larviculture hatcheries [19]. Probiotics could be used for the inhibitory studies because of its versatile nature such as inhibitory compounds production, competition for nutrients, competition for adhesion sites in the gastrointestinal tract, enhancement of the immune response, production of essential nutrients such as vitamins, fatty acids, and enzymatic contribution to digestion [20, 21]. Bacteria that are able to improve the water quality by removing toxic inorganic nitrogen or by mineralizing organic matter are also considered as probiotics. Bacterial strains dominantly present in culture water at high densities are also assumed to have the ability to compete efficiently for nutrients with possibly deleterious strains [20]. The development of drug resistant bacteria and the reduced efficiency of antibiotic resistant for human and animal diseases, have led to suggestions of the use of nonpathogenic bacteria as probiotic agents to control diseases (**Figure 3**).

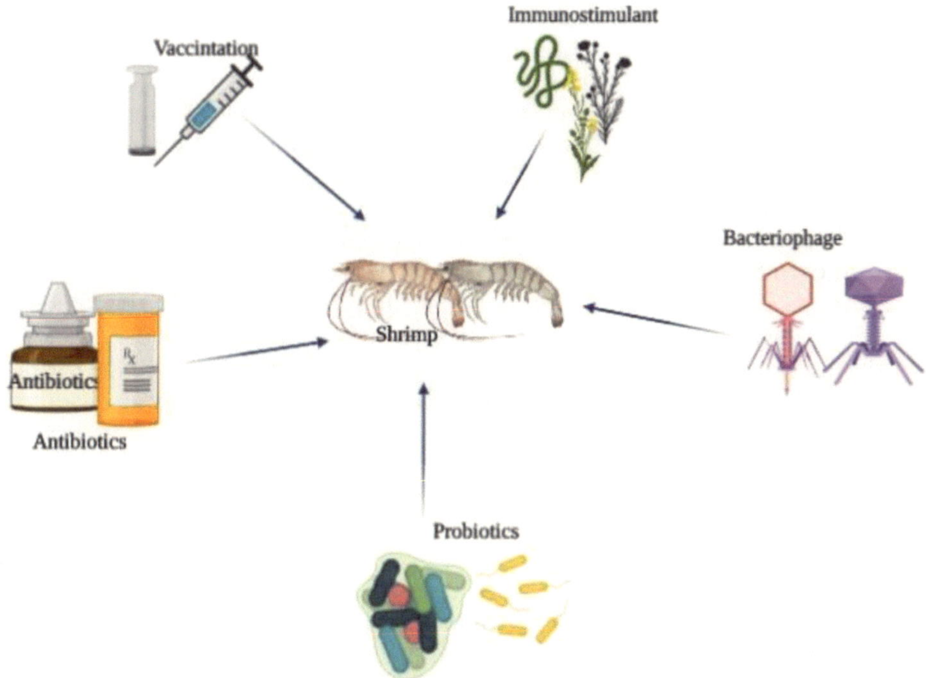

Figure 3.
Various strategies for the shrimp bacterial disease.

9. Bacteriophages

Recently, bacteriophages (phages) are proposed as candidate therapeutics for aquaculture [22]. They could reduce pathogenic bacteria safely, effectively, and ecofriendly, as they are the natural enemies of bacteria. A major advantage of phage therapy is that non–target microbiota is not affected because the phages usually have a narrow host range [18]. However, phages can transfer virulence factors rapidly and selective pressure on the *Vibrio* population might select for strains that are non–sensitive to the phage. Biofilm formation and control A biofilm is an assemblage of microbial cells that is irreversibly associated with a surface and enclosed in a matrix of primarily polysaccharide material. A biofilm is defined as a bacterial population in which the cells adhere to each other and to surfaces or interfaces with architectural complexity [23]. Bacteria are known to colonize the surfaces immediately after the formation of a film of organic molecules. A bacteria cell is grouped together and forms the microbial aggregation on the surface of various materials such as pipes and tanks and form biofilms [24]. All bacterial strains does not have the capacity to produce biofilm layers, particular group of bacterial strains has the potential activity to produce the biofilm layers. In aquaculture system the biofilms are produced by the important bacterial pathogen *Vibrio* sp., which causes large-scale mortalities in shrimp and prawn hatcheries [24]. Biofilm formation may be an important mechanism for host immune modulation and virulence factor for down regulation [25]. Bacterial proliferation in the intestines is thought to induce quorum sensing, which down regulates extracellular matrix production and adhesions required for biofilm formation [26]. The primary quorum sensing system in bacteria that appears to control important adhesins for in vitro biofilm formation is AI-3, an alternative class of quorum sensing molecule distinct from previously reported AHLs and AI-2 [27]. AI-3 also controls other virulence factors such as the locus of enterocyte effacement (LEE) pathogenicity, which encodes for type III secretion system (T3SS) and toxins secreted by T3SS. AI-3 receptor quorum sensing also binds to epinephrine and norepinephrine [28]. Very few biofilm models of animal infections have been established, and even fewer have tested virulence of quorum sensing mutants in such models. Many research focuses towards the bacterial biofilm inhibition and eradication strategies, the potential antimicrobial agent production is very crucial to treat these biofilm problems in aquaculture industry. Some of the studies already proved the cleavage of bacterial extrapolysachharide layers and successfully inhibit the bacterial biofilms; however these compounds are not work well in all bacterial biofilms. Hence, the researches towards the development the biofilm inhibition through various approaches are enhanced. The use of enzyme-based detergents as biocleaners, also known as "green chemicals," can serve as a viable option to overcome the biofilm problems [29].

10. N-acyl homoserine lactonase

The first AHL-degrading enzyme was identified from Bacillus sp. expression of its gene in *Erwinia carotovora*, pathogenicity in plants has been reported [30]. Many AHL enzymes identified in bacteria, fungi, and mammals [31]. Paraoxonases (PONs) from mammalian sera also have lactonase-like activities in addition to their involvement in the hydrolysis of organophosphates [32]. AHL-degrading enzyme, with both short and long chain AHLs as substrates and little activity with other chemicals was documented earlier [33]. Screening of bacteria capable of producing enzymes, which inactivate the signal compound, blocking the quorum sensing systems of their competitors, has potential for disease control in aquaculture [18]. Diverse aquatic bacteria employ signal molecules to regulate the production of virulence factors [34]. Disruption of these signal molecules can significantly decrease virulence factor production in bacteria without interfering with their growth and it may be a particularly useful method in aquaculture [19, 35, 36]. One of the approaches proposed for quorum sensing disruption is the isolation of bacteria that degrade signal molecules involved in quorum sensing. Bacteria capable of utilizing N–acyl homoserine lactone (AHL) molecules as sole source of carbon and nitrogen can be used as potential quenchers of quorum sensing regulated functions in pathogenic bacteria. Bacteria capable of degrading AHL–type signal molecules have been reported extensively in the literature [37]. Hence, it is of interest to investigate whether these types of bacteria could be used as a new General Introduction: a report on N-acyl homoserine lactonase from quorum quenching *Bacillus licheniformis* and its control of *Vibrio parahaemolyticus* colonization in *Fenneropenaeus indicus* type of probiotic, a live microbial adjunct that is beneficial to the host [18]. Enrichment cultures of AHL degrading bacteria controlled the overall microbial activity in aquaculture. The addition of AHL presumably stimulates the virulence of opportunistic pathogenic bacteria. Assuming that in more intensive aquaculture systems, the often observed high mortality is related to the presence of quorum sensing molecules and quorum sensing induced virulence factors, the addition of an N-acyl homoserine lactonase could be beneficial [38]. The ability to degrade AHLs is widely distributed in the bacterial kingdom, isolated from soil by enrichment culture, is able to utilize AHL compounds as sole carbon, nitrogen and energy source. Bacterial species in natural environments that can metabolize AHLs and disrupt quorum sensing regulation in nearby bacteria was indicated [39]. Bacterial species that interfere with quorum sensing regulation in another species were reported [40]. The AHL inactivation activity in *Bacillus cereus* isolate was due to its synthesis and secretion of a lactonase capable of opening the homoserine lactone ring of AHLs, thereby reducing the effectiveness of the signal molecules [41]. Some bacteria, especially Bacillus sp. may use AHL-lactonases in quorum-quenching to boost competitive strength in soil. Metallo-β-lactamase superfamily (MBL) family-type enzymes have been characterized from a variety of soil-associated *Bacillus* spp. and other bacteria [42, 43]. Metallo-lactamases consist of conserved motif HXHXDH and a zinc binding motif, and dinuclear zinc binding center bridged by an aspartate and an oxygen species [44, 45]. AHL lactonase AiiB from *Agrobacterium tumefaciens* also posses the similar active sites [46]. AHL lactonase expression in the pathogens *Erwinia amylovora*, *Pseudomonas aeruginosa* PAO1 and *Burkholderia cepacia* reduced their virulence by degrading AHLs [47]. In addition, AHL lactonases have also been expressed in *Escherichia coli* and *Pichia pastoris* [48]. Green fluorescent protein The introduction of green fluorescent protein (GFP) as an endogenous fluorescent tag provides a mean of rendering the bacteria visible and also tracing their activity in living host cells [49]. Green fluorescent protein is a small protein (27 KDa) found in the jellyfish, *Aequorea victoria*. It has the property of fluorescing when excited by ultraviolet light [50]. This gene codes for a fluorescent protein, when excited with UV light (470 nm), emits a wavelength of 502 nm. The GFP fluoresce is independent of cofactors, substrates or any additional gene products, sensitive, stable, specific, non-toxic and does not interfere with cell growth and function. GFP-marked fish pathogens have been constructed to study the invasion pathway in fish models [51]. The fate of *Vibrio parahaemolyticus* once filtered by oysters and its capacity to proliferate in different post-harvest conditions was defined, using a strain of *V. parahaemolyticus* with a plasmid that contains the GFP gene [52]. Recently, GFP was used to study colonization and pathogenesis of *Vibrio parahaemolyticus* in different tissues and hemolymph of *F. indicus* [53].

11. The prophenoloxidase activating system (proPO system)

The proPO system is an efficient part of the innate immune response and consists of several proteins, which are involved in pattern recognition proteins, proteases, protease inhibitors, antioxidants proteins and melanisation as represented in **Figure 4**.

In addition, it also associated with the cytotoxic reactions, cell adhesion encapsulation, and phagocytosis, which is present in many invertebrate groups, such as ascidians, mollusks, echinoderms, millipedes, bivalves, brachiopods and insects [54, 55]. In invertebrates the humoral mediated immune system is triggered through several hemolymph proteins amongst the prophenoloxidase plays the vital role against the invading pathogens. At the same time, this immune pathway is stimulated through microbial pathogens such as bacteria, fungi and virus. The stimuli are derived from the outer membrane components of microbes, those molecules are termed as pathogen-associated molecular pattern (PAMP) which are lipopolysaccharide (LPS) and peptidoglycans (PG) from bacteria and β-glucans from fungi. This proPO cadcade consist of pattern-recognition proteins (PRPs) including LPS and β-1,3-glucan-binding protein (LGBP), β-1,3 glucan binding protein (βGBP), and peptidoglycan binding protein (PGBP), several serine protease and zymogens, proPO as well as proteinase inhibitors, which are important regulatory factors to avoid activation of the system where it is not appropriate [56].

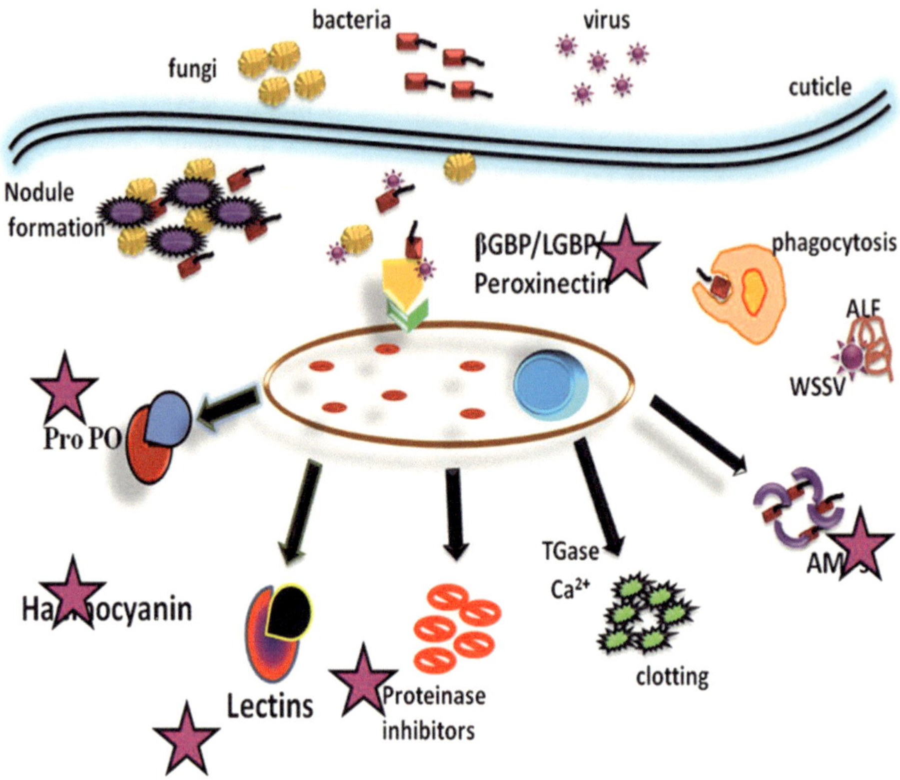

Figure 4.
Schematic outline of the principle components in the prophenoloxidase (proPO)-activating system in arthropods.

12. Protein mediated nanoparticles

Alternative approaches to treat bacterial infections are urgently needed in aquaculture worldwide. Nanobiotechnology and nanotechnology products have a wide usage potential in aquaculture and seafood industries. For instance, production of more effective fish feed for aquaculture species by the application of nanotechnology is possible. New materials obtained by the nanosciences can be used in the different aspects of fisheries and aquaculture. Nanotechnology may have the potential to provide aquaculture that is safe from disease and pollution. Use of quorum quenching enzymes as antimicrobial agents is nature-inspired and has recently attracted much attention as an antibiotic-free approach to treat bacterial infections. The use of antimicrobial enzymes covalently attached to nanoparticles is of special interest because of enhanced stability of protein-nanoparticle conjugates and the possibility of targeted delivery.

13. Antimicrobial peptides

AMPs are effectors of the innate immune system and function as a first line of defense to fight against invading microorganisms [57] are represented in **Figure 5**. Therefore, AMPs are critical for shrimp to fight against the pathogenic invasion. AMPs are typically small in size, are naturally derived or synthetic and are active against a wide range of microorganisms, such as bacteria, virus, yeast, parasite and fungi,

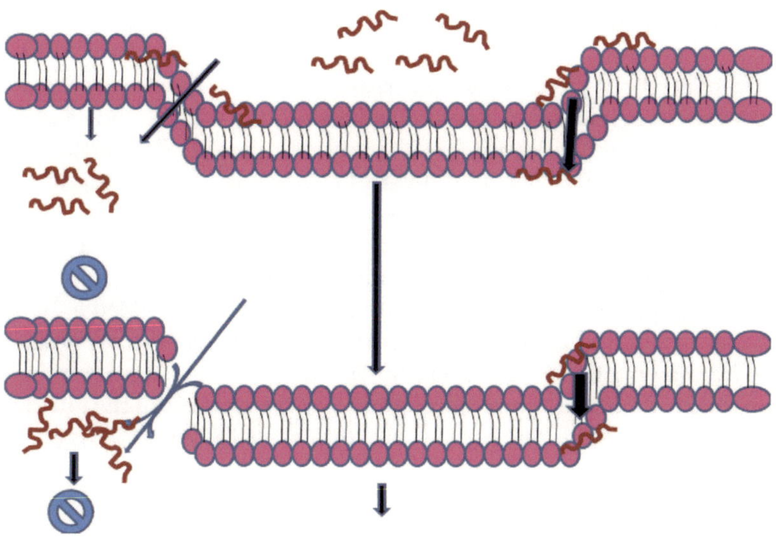

Figure 5.
Antimicrobial peptides intrusion mechanism inside the bacterial cell wall.

and they may also exhibit an anti-tumor activity [57, 58]. Generally, it has less than 150–200 amino acid residues, and it has an amphipathic structure with cationic or anionic properties. Several families of shrimp AMPs, such as penaeidins, lysozymes, crustins, ALFs and stylicins, have been identified and characterized [59, 60]. They are produced by and stored in the hemocytes; these are key cells in the crustacean immune system [61]. Various methods are discussed in the introduction section to eradicate the bacterial inhibition and bacteria causing disease management. However, in this chapter, we used the antimicrobial peptides to inhibit the bacterial causing biofilms and using probiotic bacteria, we attempted to reduce the bacterial disease.

13.1 Crustins

Crustins are generally defined as multi-domain cationic antibacterial polypeptides (7–14 kDa) containing a whey acidic protein (WAP) domain at the C-terminus (**Figure 5**) [62]. The first identified crustin member is an 11.5 kDa protein purified from the granular haemocytes of the shore crab, *Carcinus maenas* that exhibits specific activity towards Gram-positive marine or salt-tolerant bacteria [63, 64]. Over 50 crustins and crustin-like sequences have been identified in numerous crustacean species, including crayfish, shrimp, freshwater prawn, crab, lobster, and also in non-decapod crustaceans, such as amphipods, (through EST-based approaches) [65].

14. Materials and methods

14.1 Collection and maintenance of bacterial strains

Fenneropenaeus indicus, Penaeus monodon, Litopenaeus vannamei, and *Penaeus semisulcatus* were collected from different sea shore area in and around Tamil Nadu. Live *P. semisulcatus* was acclimatized in lab for two week before the experimentation. All the tanks received continuous aeration, and 50% of the water was exchanged daily to maintain quality. Pathogenic strains and probiotics bacteria were isolated from the whole intestinal tract and hepatopancreas of wild caught *P. semisulcatus* larvae.

14.2 Replica plating method

The shrimp intestinal tract, hepatopancreas content was aseptically removed from a live healthy prawn were homogenized and serially diluted with sterilized normal saline solution. Suspensions (0.1 ml) were spread on different media like nutrient agar (with 1% w/v NaCl) and thiosulfate/citrate/bile/salt (TCBS), Zobell Marine agar (ZMA), bacillus medium (HiMedia, India), and incubated under an aerobic atmosphere overnight at 37°C for 24 h. After incubation, predominant bacterial colonies were selected based on their morphological characters including color, shape, size colonies from all media were replica plated on the Muller Hinton agar medium (Hi Media, India) with target bacterial strains and incubated at room temperature for 24 h. After incubation viable colonies showing the zone of clearance against the target *Vibrio* strains were streaked on the nutrient medium to check purity of the isolate. All the purified strain was maintained in Zobell Marine Broth (HiMedia) at −20°C with 15% glycerol.

14.3 Antagonism assay

To evaluate the potential antagonistic activity of the isolated probionts by well diffusion assay on solid medium and eight *Vibrio* spp. was used for our study. *Vibrio* spp. was precultured in broth for 24 h and incubated at 28°C. Lawns were prepared by spreading 50 μl of each target strain (*Vibrio* spp.). Wells were cut in LA plates with a sterile 5-mm cork borer and filled with 50 μl (10^8 CFU) of the 72 h old each and probiotics cell free extract were carefully pipetted into each well. Two probiotics were tested per plate in triplicate. The diameter of the inhibition zones around the wells were recorded in millimeters after incubating the plates for 24 h at 25°C.

14.4 Characterization of strains

Isolated strains were subjected to standard morphological, biochemical assay followed by bacterial genomic DNA was extracted using the method [66]. Bacterial strains were cultivated in 10 mL of Luria-Bertani broth (LB) at 29°C in agitation for 18 h. culture was centrifuged for 5000 rpm for 5 min suspended pellet with Sucrose TE buffer 10 $mmol^{-1}$ lysozyme was added and incubated for 30 min. After incubation, add 100 μl of 0.5 M EDTA (pH 8) and 60 μl of 10% SDS added with 250 μl of equilibrated Phenol and 250 μl of Chloroform and mixed gently mixture was centrifuged. An equal volume of chloroform and isoamyl alcohol mixture (24:1) was added with shaking the mixture. Collect the aqueous phase in a sterile tube and precipitate it with 2 volumes of 100% ethanol and 3 M Sodium Acetate. Store at −20°C for 30 min. Followed by this addition, the sample was centrifuged with at 12000 rpm and the isolated DNA was precipitated with 70% ethanol. DNA was suspended with 30 μl of TE Buffer pH (8.0) and DNA was extracted for 16S rRNA sequence determination & RAPD analysis.

14.5 rRNA gene amplification

The region of the 16S ribosomal gene (rRNA) of the DNA extracted from each bacterial strain was amplified by the polymerase chain reaction (PCR). The reagent mixture was prepared with the universal 16S rRNA Fp 5′-AGA GTT TGA TCC TGG CTC AG-3′ and 16S rRNA Rp 5′-ACG GCT ACC TTG TTA CGA CTT-3′ [67], samples were amplified by PCR in Std buffer 2.5 μl, dNTPs 0.5 μl, forward and reverse primers each 1.0 μl and Taq 0.2 μl and template DNA 1.0 μl condition consist of 40 cycles of 95°C (5 min), 55°C (1 min), and 72°C for (2 min) and with final 72°C for 10 min for elongation process were performed with four bacteria strains, yielding positive amplification for all DNA tested, as determined by visualization on agarose gel electrophoresis. The amplification products were purified by using Real genomics kit, by following the specifications of the manufacturer.

14.6 Co-culture method

The co-culture method was performed to observe the antagonistic potential and reproductive effect of the isolated bacteria, when grown with the *Vibrio* spp. in a same medium. Culture broth of the prospective probiont and the target organism were inoculated into LB broth to check the probiotic activity of our isolated culture. Two Probiotic strains (DMP4 & DMB3) were used. Eight selected *Vibrio* spp., (v1, v2, v3, v4, v5, v6, v7, v8) visibly different from each other in size, shape and color of the colony morphology were used for co-culture method. The initial cell density of selected *Vibrio* strains was approximately 10^3 CFU ml^{-1}, whereas the initial concentration of probiotic cell free extract was 10^5, 10^6, 10^7, and 10^8 CFU ml^{-1}. All inoculums were prepared with LB broth each *Vibrio* inoculum and each probiotic inoculum was mixed together, and then incubated for 24 h days at 25°C and reading was taken in every 2 h intervals.

14.7 Artemia hatchability test

Probiotic activity of chosen strains were further tested with Artemia hatchability test, 160 mg of dried cysts were hatched in 80 ml of sterile sea water under the conditions of strong aeration and constant illumination, at 28°C *Vibrio* sp. was grown in leuria broth incorporated with NaCl enrichment (3%) and up to 10^8 CFU ml^{-1} was obtained within 24 h. The concentration of *Vibrio* sp. was dispensed in 10^3 CFU ml^{-1} in hatching unit. Six experimental group along with one control without any pathogen or probiotic bacteria was subjected for *hatchability test*. Five experimental groups namely *Vibrio* sp. + *Pseudomonas* sp. (vp), *Vibrio* sp. + *Bacillus* sp. (vb).positive control group *Bacillus* sp. (vb) (dahb) and *Pseudomonas* sp.(vp) (dahp) were taken for the experiment pathogenic *Vibrio* sp. was inoculated at conc. 10^3 CFU ml^{-1} and Probiotics bacteria dahp and dahb were inoculated in 10^8 CFU ml^{-1} to the corresponding experimental set up along with prehydrated cyst the hatchability was assessed for 24 h [68]. After 24 h exposure, the free nauplii were counted under microscope. Five replicates were calculated for the control and each treatment measures. The hatching percentage (%h) was calculated with following formula:

$$\%H = \frac{N}{N + C + U} \times 100 \tag{1}$$

where N = Nauplii, C = Unhatched cysts, and U = Umbrella stage.

14.8 Challenge studies in *Artemia nauplii*

Five experimental groups namely *Vibrio* sp. + *Pseudomonas* sp. (Vp), *Vibrio* sp. + *Bacillus* sp. (Vb).positive control group *Bacillus* sp. (b) and *Pseudomonas* sp. (p) were taken for the experiment *Vibrio* sp. was inoculated at conc. 10^3 CFU ml^{-1} and Probiotics bacteria *Pseudomonas* sp. and *Bacillus* sp. (p&b)

were inoculated in 10^8 CFU ml^{-1} The survival rate was determined in every 2 h intervals of exposure for 12 h. The percentage of survival was calculated by the formula

$$\text{Survival rate}(\%) = \frac{\text{Number of live nauplii } \textit{at} \text{ every 2 h interval}}{\text{Number of nauplii } \textit{at} \text{ the time of inoculation}} \quad (2)$$

The active nauplii were considered as live and counted under microscope.

15. Discussion

Various approaches are applied to eradicate the bacterial and other microbial diseases in aquaculture, however this chapter deals with the biofilm inhibitory mechanism through two different ways probiotic isolation from aquaculture farms or the specific animals such as shrimp or crab form the body part. For this many microbiology techniques are used to isolate the bacteria and identify, then inhibit the *Vibrio* causing biofilm in these probiotic strains. Another approach of treating bacterial biofilms via antimicrobial peptides that also derived from marine sources could be used here. Such kinds of peptides easily penetrates into the layer of bacterial cell membranes and quickly disrupt and dissolve the polysaccharide layer, which leads to the inhibition of biofilms in the surface of the layer.

16. The potential use of antimicrobial peptides for disease control in aquaculture

Antimicrobial peptides provide a good therapeutic alternative for the treatment of diseases in aquaculture. Several antimicrobial peptides from various sources are already in clinical and commercial use [69]. However, it is quite promising that the shrimp AMPs could be potential candidates as an alternative to antibiotics in shrimp farming. Besides their antimicrobial function, AMPs are also known to act as mediators of inflammation influencing diverse processes such as cell proliferation, wound healing, cytokine release and immune induction [70].

The significance of aquaculture in the context of global food production sector, the management of aquatic resources and the socio-economic development of coastal rural areas is now fully appreciated world-wide. In the last decade, a series of papers describing shrimp immunity were published and a batch of related data accumulated, which are very useful for understanding the interaction between shrimp and pathogens to enrich the immune theory of invertebrates. Recently, several review papers summarize the achievements in shrimp immunity including expressed sequenced tag and database construction [71], microarray analysis of shrimp immune response [72], shrimp molecular responses to viral pathogen [73] and the cationic antimicrobial peptides in penaeid shrimp [60]. Obviously understanding the shrimp immunology is necessary to develop an effective strategy for disease control. Indian white shrimp *Fenneropenaeus indicus* is an important crustacean for aquaculture and has brought significant revenue to rural economy. The epidemics, however, caused a dramatic mortality and resulted in a severe economic loss. Therefore, the understanding of the immune ability of shrimp and their defense mechanisms has become a primary concern in shrimp aquaculture.

17. Conclusions

Recent advances gives the notions to incorporate the different techniques and utilized in the aquaculture practical system as a combined mode to enhance the potential activity of the strategies towards the microbial pathogens, this would be efficient method when compare to conventional techniques alone. In addition this way of microbial eradication will helps to improve the aquaculture production as well as cost effective.

Acknowledgements

The study was supported by the National Research Foundation of Korea, which is funded by the Korean Government [NRF-2018-R1A6A1A-03024314].

Conflict of interest

The authors declare no conflict of interest.

Author details

Jeyachandran Sivakamavalli[1*], Kiyun Park[1], Ihn–Sil Kwak[1,2]
and Vaseeharan Baskaralingam[3]

1 Fisheries Science Institute, Chonnam National University, Yeosu, South Korea

2 Faculty of Marine Technology, Chonnam National University, Yeosu, South Korea

3 Department of Animal Health and Management, Alagappa University, Karaikudi, India

*Address all correspondence to: vallipdf@gmail.com

References

[1] Flegel TW, Lightner DV, Lo CF, Owens L. Shrimp disease control, past, present and future. In: Bondad-Reantaso MG, Mohan CV, Crumlish M, Subasinghe RP, editors. Diseases in Asian Aquaculture VI. Manila, Philippines: Fish Health Section, Asian Fisheries Society; 2008. pp. 355-378

[2] Loof TG, Schmidt O, Herwald H, Theopold U. Coagulation systems of invertebrates and vertebrates and their roles in innate immunity: The same side of two coins? Journal of Innate Immunity [Internet]. 2011;**3**(1):34-40. DOI: 10.1159/000321641

[3] Jiravanichpaisal P, Lee BL, Söderhäll K. Cell-mediated immunity in arthropods: Hematopoiesis, coagulation, melanization and opsonization. Immunobiology. Elsevier BV [Internet]. 2006 Jun;**211**(4):213-236. DOI: 10.1016/j.imbio.2005.10.015

[4] Soderhall K. Invertebrate immunity. Developmental & Comparative Immunology. 1999;**23**:263-266

[5] Theopold U, Li D, Fabbri M, Scherfer C, Schmidt O. The coagulation of insect hemolymph. Cellular and Molecular Life Sciences, CMLS [Internet]. 2002;**59**(2):363-372. DOI: 10.1007/s00018-002-8428-4

[6] Olafsen. Interactions between fish larvae and bacteria in marine aquaculture. Aquaculture. 2001;**200**(1-2):223-247

[7] Russell AD. Biocide use and antibiotic resistance, the relevance of laboratory findings to clinical and environmental situations. The Lancet Infectious Diseases. 2003;**3**(12):794-803

[8] Sorum H. Antimicrobial drug resistance in fish pathogens. In: Aarestrup FM, editor. Antimicrobial Resistance in Bacteria of Animal Origin. Washington DC: American Society for Microbiology Press; 2006. pp. 213-238

[9] Watson AK, Heinrich Kaspar M, Josie L, Lewis G. Probiotics in aquaculture, the need, principles and mechanisms of action and screening processes. Aquaculture. 2008;**274**(1):1-14

[10] Selvin J, Lipton AP. Leaching and residual kinetics of oxytetracycline incorporated medicated feed treated to juvenile black tiger shrimp *Penaeus monodon* fabricious. Fishery Technology. 2009;**41**(2):93-100

[11] Vaseeharan B, Ramasamy P, Chen JC. Antibacterial activity of silver nanoparticles (AgNps) synthesized by tea leaf extracts against pathogenic *Vibrio harveyi* and its protective efficacy on juvenile *Feneropenaeus indicus*. Letters in Applied Microbiology. 2010;**50**(4):352-356

[12] ImmanuelG,SivagnanavelmuruganM, Balasubramanian V, Palavesam A. Effect of hot water extracts of brown seaweeds *Sargassum* spp. on growth and resistance to white spot syndrome virus in shrimp *Penaeus monodon* postlarvae. Aquaculture Research. 2010;**41**(10):545-553

[13] Castro SBR, Leal CAG, Freire FR, Carvalho DA, Oliveira DF, Figueiredo HCP. Antibacterial activity of plant extracts from Brazil against fish pathogenic bacteria. Brazilian Journal of Microbiology. 2008;**39**(4):756-760

[14] Pandey G, Madhuri S, Mandloi AK. Medicinal plants useful in fish diseases. Plant Archives. 2012;**12**(1):1-4

[15] Gudding R, Lillehaug A, Evensen O. Recent developments in fish vaccinology. Veterinary Immunology and Immunopathology. 1999;**72**(1-2):203-212

[16] Thanardkit P, Khunrae P, Suphantharika M, Verduyn C. Glucan from spent brewer's yeast, preparation, analysis and use as a potential immunostimulant in shrimp feed. World Journal of Microbiology and Biotechnology. 2002;**18**(6):527-539

[17] Marques A, Dinh T, Ioakeimidis C, Huys G, Swings J, Verstraete W, et al. Effects of bacteria on *Artemia* cultured in different gnotobiotic environments. Applied and Environmental Microbiology. 2005;**71**(8):4307-4317

[18] Defoirdt T, Boon N, Sorgeloos P, Verstraete W, Bossier P. Alternatives to antibiotics to control bacterial infections, luminescent vibriosis in aquaculture as an example. Trends in Biotechnology. 2007;**25**(10):472-479

[19] Tinh NT, Dierckens K, Sorgeloos P, Bossier P. A review of the functionality of probiotics in the larviculture food chain. Marine Biotechnology. 2008;**10**(1):1-12

[20] Verschuere L, Heang H, Criel G, Sorgeloos P, Verstraete W. Selected bacterial strains protect *Artemia* spp. from the pathogenic effects of *Vibrio proteolyticus* CW8T2. Applied and Environmental Microbiology. 2000;**66**(3):1139-1146

[21] Vine NG, Leukes WD, Kaiser H. Probiotics in marine larviculture. FEMS Microbial Reviews. 2006;**30**(3):404-427

[22] Nakai T, Park SC. Bacteriophage therapy of infectious diseases in aquaculture. Research in Microbiology. 2002;**153**(1):13-18

[23] Lawrence JR, Horber DR, Hoyle BD, Costerton JW, Caldwell DE. Optical sectioning of microbial biofilms. Journal of Bacteriology. 1991;**173**(20):6558-6567

[24] Karunasagar I, Otta SK, Karunasagar I. Biofilm formation by *Vibrio harveyi* on various surfaces. Aquaculture. 1996;**140**(3):241-245

[25] Irie T, Watarai S, Iwasaki T. Binding of *Vibrio anguillarum* to neutral glycosphingolipids from intestinal mucosa of rainbow trout (*Oncorhynchus mykiss*). Veterinary Medicine and Science. 2004;**66**(2):205-208

[26] Hammer BK, Bassler BL. Quorum sensing controls biofilm formation in *Vibrio cholera*. Molecular Microbiology. 2003;**50**(1):101-114

[27] Sperandio V, Torres AG, Jarvis B, Nataro JP, Kaper JB. Bacteria-host communication, the language of hormones. Proceedings of the National Academy of Sciences of the United States of America. 2003;**100**(15):8951-8956

[28] Clarke MB, Hughes DT, Zhu C, Boedeker EC, Sperandio V. The QseC sensor kinase, a bacterial adrenergic receptor. Proceedings of the National Academy of Sciences of the United States of America. 2006;**103**(27):10420-10425

[29] Meyer B. Approaches to prevention, removal and killing of biofilms. International Biodeterioration & Biodegradation. 2003;**51**(4):249-253

[30] Dong YH, Xu JL, Li XZ, Zhang LH. AiiA, an enzyme that inactivates the acylhomoserine lactone quorum sensing signal and attenuates the virulence of *Erwinia carotovora*. Proceedings of the National Academy of Sciences. 2000;**97**(7):3526-3531

[31] Czajkowski R, Jafr S. Quenching of acyl-homoserine lactonedependent quorum sensing by enzymatic disruption of signal molecules. Acta Biochimica Polonica. 2009;**56**(1):1-16

[32] Yang F, Wang LH, Wang J, Dong YH, Hu JY, Zhang LH. Quorum quenching enzyme activity is widely conserved in the sera of mammalian species. FEMS Letters. 2005;**579**(17):3713-3717

[33] Wang LH, Weng LX, Dong YH, Zhang LH. Specificity and enzyme kinetics of the quorum-quenching *N*-acyl homoserine lactone lactonase (AHL lactonase). The Journal of Biological Chemistry. 2004;**279**(14):13645-13651

[34] Baruah K, Dang TV, Cam KD, Mathieu W, Defoirdt T, Sorgeloos P, et al. In vivo effects of single or combined N-acyl homoserine lactone quorum sensing signals on the performance of *Macrobrachium rosenbergii* larvae. Aquaculture. 2009;**288**(1):233-238

[35] Finch RG, Pritchard DI, Bycroft BW, Williams P, Stewart GSAB. Quorum sensing, a novel target for anti-infective therapy. Antimicrobial Chemotherapy. 1998;**42**(5):569-571

[36] Hentzer M, Wu H, Andersen JB. Attenuation of *Pseudomonas aeruginosa* virulence by quorum sensing inhibitors. EMBO Journal. 2003;**22**(15):3803-3815

[37] Uroz S, D'Angelo-Picard C, Carlier A, Elasri M, Sicot C, Petit A, et al. Novel bacteria degrading N-acylhomoserine lactones and their use as quenchers of quorum-sensing-regulated functions of plant-pathogenic bacteria. Microbiology. 2003;**149**(8):1981-1989

[38] Fast W, Tipton PA. The enzymes of bacterial censusand censorship. Trends in Biochemical Sciences. 2012;**37**(1):7-14

[39] Leadbetter JR, Greenberg EP. Metabolism of acyl homoserine lactone quorum-sensing signals by *Variovorax paradoxus*. Journal of Bacteriology. 2000;**182**(24):6921-6926

[40] Dong YH, Zhang LH. Quorum sensing and quorum quenching enzymes. Microbiology. 2005;**43**(1):101-109

[41] Bauer WD, Robinson JB. Disruption of bacterial quorum sensing by other organisms. Current Opinion in Biotechnology. 2002;**13**(3):234-237

[42] Rasmussen TB, Manefield M, Andersen JB, Eberl L, Anthoni U, Christophersen C, et al. How *Delisea pulchra* furanones affect quorum sensing and swarming motility in *Serratia liquefaciens* MG1. Microbiology. 2000;**146**(12):3237-3244

[43] Roche DM. Communications blackout? Do N-acyl homoserine lactone-degrading enzymes have any role in quorum sensing? Microbiology. 2004;**150**(7):2023-2028

[44] Momb J, Thomas PW, Breece RM, Tierney DL, Fast W. The quorum-quenching metallo-β-lactonase from *Bacillus thuringiensis* exhibits a leaving group thio effect. Biochemistry. 2006;**45**(44):13385-13393

[45] Liu D, Lepore B, Petsko W, Thomas GA, Stone PW, Fast EM, et al. Three-dimensional structure of the quorum-quenching N-acyl homoserine lactone hydrolase from *Bacillus thuringiensis*. Proceedings of the National Academy of Sciences of the United States of America. 2005;**102**(33):11882-11887

[46] Cao Y, He S, Zhou Z, Zhang M, Mao W, Zhang H, et al. Orally administered thermostable *N*-Acyl Homoserine Lactonase from *Bacillus* sp. strain AI96 attenuates *Aeromonas hydrophila* infection in Zebrafish. Applied and Environmental Microbiology. 2012;**78**(6):899-1908

[47] Molina L, Constantinescu F, Michel L, Reimmann C, Duffy B, Defago G. Degradation of pathogen quorum-sensing molecules by soil bacteria, a preventive and curative biological control mechanism. FEMS Microbiology Ecology. 2003;**45**(1):71-81

[48] Chen R, Zhou Z, Cao Y, Bai Y, Yao B. High yield expression of an AHL

lactonase from *Bacillus* sp. B546 in *Pichia pastoris* and its application to reduce *Aeromonas hydrophila* mortality in aquaculture. Microbial Cell Factories. 2010;**9**(39):1475-2859

[49] Yin JF, Zhou XY, Li Q, Li Hou YH, Zhang HL. A study on the survival dynamics of *Bacillus subtilis* in water using green fluorescent protein labeling. Chinese Journal of Analytical Chemistry. 2007;**35**(10):1405-1409

[50] Chalfie M, Tu Y, Euskirchen G, WardWW, PrasherDC. Greenfluorescent protein as a marker for gene expression. Science. 1994;**263**(5148):802-805

[51] O' Toole R, Hofsten JV, Rosqvist R, Olsson PE, Wolf-Watz H. Visualisation of zebra fish infection by GFP-labelled *Vibrio anguillarum*. Microbial Pathogenesis. 2004;**37**(1):41-46

[52] Lowder M, Unge A, Maraha N, Jansson JK, Swiggett J, Oliver JD. Effect of starvation and the viable-but-non culturable state on green fluorescent protein (GFP) fluorescence in GFP-Tagged *Pseudomonas fluorescens* A506. Applied and Environmental Microbiology. 2000;**66**(8):3160-3165

[53] Vinoj G, Vaseeharan B, Brennan G. Green fluorescent protein visualization of *Vibrio parahaemolyticus* infections in Indian white shrimp *Fenneropenaeus indicus* (H Milne Edwards). Aquaculture Research. 2014;**45**(12):1989-1999

[54] Söderhäll K, Cerenius L, Johansson MW. The prophenoloxidase activating system and its role in invertebrate defence. Annals of the New York Academy of Sciences. 1994;**712**:155-161

[55] Cannon B. Venezuela, April 2002: Coup or popular rebellion? The myth of a united Venezuela. Bulletin of Latin American Research. 2004;**23**(3):285-302

[56] Cerenius L, Söderhäll K. The prophenoloxidase-activating system in invertebrates. Immunological Reviews. 2004;**198**(1):116-126

[57] Hancock RE, Diamond G. The role of cationic antimicrobial peptides in innate host defences. Trends in Microbiology. 2000;**8**(9):402-410

[58] Krepstakies M, Lucifora J, Nagel CH, Zeisel MB, Holstermann B, Hohenberg H, et al. A new class of synthetic peptide inhibitors blocks attachment and entry of human pathogenic viruses. The Journal of Infectious Diseases. 2012;**205**(11):1654-1664

[59] Rolland JL. Aspects moléculaires et biochimiques des stylicines, peptides multifonctionnels identifiés chez la crevette bleue du Pacifique Litopenaeus stylirostris (Crustacea, Decapoda) [Doctoral dissertation]. Université Montpellier 2-Sciences et Techniques du Languedoc; 2010

[60] Tassanakajon A, Amparyup P, Somboonwiwat K, Supungul P. Cationic antimicrobial peptides in penaeid shrimp. Marine Biotechnology. 2010;**12**(5):487-505

[61] Somboonwiwat K, Marcos M, Tassanakajon A, Klinbunga S, Aumelas A, Romestand B, et al. Recombinant expression and anti-microbial activity of anti-lipopolysaccharide factor (ALF) from the black tiger shrimp *Penaeus monodon*. Developmental & Comparative Immunology. 2005;**29**(10):841-851

[62] Smith VJ, Fernandes JM, Kemp GD, Hauton C. Crustins: Enigmatic WAP domain-containing antibacterial proteins from crustaceans. Developmental & Comparative Immunology. 2008;**32**(7):758-772

[63] Relf JJ, Chisholm GK, Smith V. Purification and characterization of a cysteine-rich 11.5-kDa antibacterial protein from the granular haemocytes

of the shore crab, *Carcinus maenas*. European Journal of Biochemistry. 1999;**264**:350-357

[64] Kalishwaralal K, Barath MS, Ramkumarpandian S, Deepak V, Gurunathan S. Silver nanoparticles impede the biofilm formation by *Pseudomonas aeruginosa and Staphylococcus epidermidis*. Colloids and Surfaces B: Biointerfaces. 2010;**79**(2):340-344

[65] Sun L. Peptide-based drug development. Modern Chemistry and Applications. 2013;**1**(1):1-2

[66] Sambrook J, Fritschi EF, Maniatis T. Molecular cloning: a laboratory manual. New York: Cold Spring Harbor Laboratory Press; 1989

[67] Weisburg WG, Barns SM, Pelletier DA, Lane DJ. 16S ribosomal DNA amplification for phylogenetic study. Journal of Bacteriology [Internet]. 1991;**173**(2):697-703. DOI: 10.1128/jb.173.2.697-703.1991

[68] Sorgeloos P, Dhert P, Candreva P. Use of the brine shrimp, *Artemia* spp., in marine fish larviculture. Aquaculture. 2001;**200**:147-159

[69] Reddy KVR, Yedery RD, Aranha C. Antimicrobial peptides: Premises and promises. International. Journal of Antimicrobial Agents [Internet]. 2004;**24**(6):536-547. DOI: 10.1016/j.ijantimicag.2004.09.005

[70] Zaiou M. Multifunctional antimicrobial peptides: therapeutic targets in several human diseases. Journal of Molecular Medicine [Internet]. 2007;**85**(4):317-329. DOI: 10.1007/s00109-006-0143-4

[71] Leu J-H, Chen S-H, Wang Y-B, Chen Y-C, Su S-Y, Lin C-Y, et al. A review of the major penaeid shrimp EST studies and the construction of a shrimp transcriptome database based on the ESTs from four penaeid shrimp. Marine Biotechnology [Internet]. 2010;**13**(4):608-621. DOI: 10.1007/s10126-010-9286-y

[72] Aoki T, Wang H-C, Unajak S, Santos MD, Kondo H, Hirono I. Microarray analyses of shrimp immune responses. Marine Biotechnology [Internet]. 2010;**13**(4):629-638. DOI: 10.1007/s10126-010-9291-1

[73] Flegel TW, Sritunyalucksana K. Shrimp molecular responses to viral pathogens. Marine Biotechnology [Internet]. 2010;**13**(4):587-607. DOI: 10.1007/s10126-010-9287-x

Chapter 5

Ants as Indicators of Terrestrial Ecosystem Rehabilitation Processes

Hendrik Sithole and Nolubabalo Tantsi

Abstract

Habitat transformation is one of the main drivers of the ecosystem degradation on earth that is ameliorated by restoring some of the degraded ecosystems by regaining their natural ecological functions with all their biotic and abiotic components. The biotic and abiotic components of the ecosystem under restoration can be used to assess the response of the ecosystem to the restoration. Ideal variable to use as the indicator should be able respond positively to the diminishing elements that we causing the degradation and interact positively to some of the biotic and abiotic components expected to prevail when the ecosystem is fully restored. One of such variable is ants. We here provide the information about the eligibility of using ants as indicators of terrestrial ecosystems undergoing restoration and sampling and basic analytical methods to apply when implanting ants at assessing ecosystem undergoing restoration.

Keywords: habitat degradation, restoration, indicator, ant sampling techniques, ant species estimation, ant species richness, ant abundance

1. Introduction

The ecological integrities of many ecosystems are presently at risk of degradation due to habitat transformation caused by several drivers including climate change [1], introductions of invasive species [2, 3], desertification, mining and heavy grazing [4, 5]. Such disturbances decrease species diversity and cause subsequent declines in ecological function and resilience in the affected ecosystems [6, 7]. Therefore, these anthropogenic disturbances or degradations should be mitigated and ecological processes restored to reduce impacts on the biological integrity of ecosystems.

The ecosystem degradation referred here is the anthropogenic disturbance or unnatural change in an ecosystem that are mainly caused by humans [8–11]. Ecosystem degradations have affected about 25% of the earth's land [9, 11], that include almost 33% of the land within Protected Areas [12, 13]. Considering that the long-term objective of the Protected Areas is to conserve nature and its associated ecosystem services and cultural values [14], such degradations within and beyond the Protected Areas threaten the achievement of this objective. The ecosystem degradation further delays the achievement of some of the Aichi Biodiversity Targets that aim to reduce and reverse the biodiversity loss on earth [15].

Ecosystem degradations can be ameliorated by implementing ecological restorations which are approaches/processes that stop and reverse the degradation to regain

 IntechOpen

the ecological functionality of the affected ecosystems – for the benefit of both nature and humans [8, 10, 16–18]. Ecosystem restoration is becoming central to conserving biodiversity and stabilizing the climate, and has started to feature prominently in global and national policy frameworks [18, 19]. It has also been recognised that ecosystem restoration will assist to achieve the 15th goal of the UNs Sustainable Development Goals that intends to protect, preserve and sustainably use the terrestrial ecosystems [20, 21], and further assists in achieving three of the Aichi Biodiversity Target in Protected areas [10] which aim to restore about 350 million hectares of degraded land globally by 2030 [22]. Furthermore the years 2021 to 2030 have been declared as decade for ecosystem restoration by the United Nations General Assembly [23].

The success rate of the ecological restoration is influenced by various factors that include the intensity and the type of the degrading agents in the system, the conditions of supporting variables such as temperature and nutrients, and the distance of the degraded system from sources that will augment its biodiversity and ecosystem services. Ecosystems that recover slow often have experienced intense degradations, or have low supporting variables, or are remote from places that can augment them with biodiversity or ecosystem services [9, 23]. These turn to cause the restoration of such ecosystems to be partially restored relative to their natural states [8, 18].

Basic measures that should be taken to restore the degraded ecosystems include the reduction or removal of pressures causing the degradation [9, 18], and then allow the natural recovery of the system (also known as passive restoration), or take further interventions (also known as active restoration) such as reintroducing/augmenting the affected species [9, 10, 24]. The restoration should include research and monitoring activities that will provide the baseline and long-term information about the progress of the restoration in achieving its goals [10, 18]. These include baseline data and information about variables related to biodiversity, trophic structures and biophysical features of the degraded area (relative to undisturbed ecosystems) that will indicate/describe the ecological integrity of the degraded area and the progress of the restoration [10, 18]. If the pre-degradation data/information is available, it can be used as the reference information to determine the extent of the degradation and the progress of restoration [18].

2. Indicators of ecosystem restoration

The ultimate aim of restoring the ecosystem is to regain its natural ecological functions where the roles of all biotic and abiotic components exist and interact naturally – without human assistance [8]. The presence of ecological variables that are facilitating these ecological functions in the recovering systems are considered enough to demonstrate that the ecosystem is being restored, or undergoing restoration [24, 25]. Their presence in an ecosystem firstly demonstrate that the factors that were degrading the system are no longer present, or are at the levels that are no longer impeding the ecological functions of the system. Secondly they demonstrate the potential of the ecological functions to occur in that ecosystem [26] for example the presence of pollination agents. In contrast, the absence of such indicators implies that the system is still not hospitable for such functions (e.g. the humus-feeding termites that were not present at the rehabilitated sites because the trees which are their primary food have not yet grown [27]), or the degrading factors are still impeding the processes of such functions to occur (e.g. deposited nitrogen hindering butterflies to colonise areas that high concentrations of it [28]).

These variables are selected to diagnose the condition or the status of the ecosystem studied. They normally comprise organisms of the ecosystem under study (mostly vegetation and animals), characteristics of the landscapes (such as the patchiness of the vegetation) and properties of the physical factors of the ecosystem (such as soil) [25]. These indicators can be directly measured (e.g. by analysing the organisms constituting the diversity in that ecosystem or measuring the size of the bare-ground) or indirectly measured where the agents associated with the concerned ecological function are measured (e.g. the concentration of some chemical elements of the soil are used to measure the quality of such soil during rehabilitation) [8, 24, 29].

3. Ants as indicators of restoration

The significant increase in the number of environmental disturbances has given rise to a need for further research to quantify the ecological effects of environmental change. Due to complexity of most ecological systems, individual species or functionally similar species groups are often used as bio-indicators of environmental processes [30]. Surveys of such indicator species help guide land managers and decision makers to identify environmental disturbances and subsequently to take actions in time to reduce damages, mitigate consequences and restore ecosystems.

One group of organisms that have been recognised as good indicators of terrestrial ecosystems are ants. Ants started been used as indicators of ecological systems in Australia in the mid 1970's assessing impacts of mining [31] and are currently used internationally assessing different degradations in many ecosystems. They were used in a five year study on a spillage pollution at the riparian of Guadiamar River, Spain, to determine their response on the areas undergoing restoration [32]. The study discovered that ants clearly respond to the restoration with species richness significantly increasing throughout this five year period, and a progressive variation in the species composition of ant communities at different riparian habitats. Another study included ants in investigating the rehabilitated sites from coal mining in Colombia and discovered two ant species (viz. *Ectatomma ruidum* and *Pheidole fallax*) could be contributing in seed dispersal and re-establishment of vegetation in these areas [33]. Another study on sites with different ages of rehabilitation from coal mining found that the nest density of ant *Pheidole fallax* ranged increases with the rehabilitation age of the sites [34]. A study on clear cut logging of timer trees in USA found that ant species assemblages respond to alteration of habitats where the populations of the invasive ant *Solenopsis invicta* and *Pheidole* species increased while the populations of the native ant species are significantly fell [35]. Another study done by Andersen and Sparling [36] found that ant species richness from site undergoing rehabilitation in Kakadu Australia, was positively correlating with the belowground soil microbial biomass which further demonstrates that the aboveground ant activities at restoration habitats can even indicate the conditions of the organisms associated with decomposition processes that are underground of these habitats.

The interactions of ants with both abiotic and biotic factors of their ecosystems make ant one of the most suitable components of the ecosystem to include when studying the impact of the degradation to the ecosystem and response of that ecosystem when undergoing/undergone restoration. One of the variables that ant can be used for as indicators, is their nests. Ants build their nests on and in variety of components within their ecosystems ranging from the soil to a specific plant species [37] which make their nest alone quite accessible to use as indicators. Some species such as *Atta bisphaerica* build their nests in the soil [38, 39], while others build theirs on or inside the plants, notably the trees like *Colobopsis nipponicus* [40, 41]. Other ant species like *Polyrhachis* species of Malaysia, build theirs on variety of objects ranging from soil to constructing them with dead vegetative and soil particles [42]. The presence or absence of the ant nests, in their habitats, could indicate the conditions of their habitats. For example Díaz [43] found no nests of ants *Messor capitatus*,

M. barbarous, *M. bouvieri* and *M. structor* built in the soil of the ploughed fields and attributed this mainly to the frequent tilling that impedes the ants to build their nests especially for winter survivorship. On other hand Sorvari and Hakkarainen [44] found high nest abandonments by ant *Formica aquilonia* at forest habitats with clear-cut logging relative to the intact habitats that they attributed to changed abiotic conditions, resource limitation, and the disturbance of the ant reproduction in clear-cut sites. These demonstrate that nests of ants can assist in indicating the conditions of the ecosystem in their habitats.

Ant species at their nest at Mokala National Park, South Africa.

The symbioses some ants have with other organisms also make them suitable as indicators of restoration. The relationship some ant species have with other organisms include hosting these organisms in their nests in exchange of some benefit from these organisms. Ant species such as *Pheidole pallidula* hosts beetle *Paussus favieri* in its nest in exchange of consuming the secretion from the beetle [45]. On other hand ant *Azteca pittier* defends the Spanish elm (*Cordia alliodora*) against browsers in exchange of getting nesting place and honeydew from the tree [46]. Some ant species even assist other organisms to supress the organisms deemed undesirable to these ants. For instance, ants genera from *Trachymyrmex* and *Acromyrmex* transport filamentous bacteria (actinomycetes) that suppress fungus *Escovopsis weberi* (the parasite of the fungi the ants cultivate) [47]. Others influence the reproduction behaviour of other organisms. For example, butterfly *Jalmenus evagoras* prefer laying her eggs on plants that have *Iridomyrmex* ant [48]. These demonstrate that apart from indicating the habitat conditions for their own, ants could indirectly indicate the conditions of other organisms they have symbiosis with. Although Hazra [49] found that ant *Pseudomyrmex ferrugineus* did not occupy *Vachellia cornigera*, for their mutualism relationship to occur, at the restored tropical forest of Mexico, the occupation of *P. ferrugineus* at tallest *V. cornigera* could be indicating that the interaction between these would commence

Author, Year & Publication	Title	Summary of the study & findings	Ant Role Studies or Discovered
Del Toro, I., Ribbons, R.R. & Pelini, S.L. 2012. *Myrmecological News* **12** pp. 133–146.	The little things that run the world revisited: a review of ant-mediated ecosystem services and disservices (Hymenoptera: Formicidae)	• Summarizes the information about ecosystem services provided by ants. • Present negative roles of ants on human and environment.	Ecosystem services
Lester, P.J., Baring, C.W., Longson, C.G. & Hartley, S. 2003. *New Zealand Entomologist* **26**: pp. 79–89.	Argentine and other ants (Hymenoptera: Formicidae) in New Zealand horticultural ecosystems: distribution, hemipteran hosts, and review	• Fifteen hemipteran species of horticultural crops in New Zealand are tended by invasive ant *Linepithema humile* and other six ant species.	Symbiosis
Canner, J.E., Dunn, R.R., Giladi, I. & Gross, K. 2012. *Acta Oecologica* **40** pp. 31–39	Redispersal of seeds by a keystone ant augments the spread of common wildflowers	• Ant *Aphaenogaster rudis* re-dispersed >90% of the seeds it took into its nest.	Seed dispersal
Chan, K.H., & Guénard, B. 2020. *Urban Ecosyst* **23** pp. 1–12	Ecological and socio-economic impacts of the red import fire ant, *Solenopsis invicta* (Hymenoptera: Formicidae), on urban agricultural ecosystems	• Invasive ant *Solenopsis invicta* decreased ant species richness and evenness in Hong Kong and impacted about 80% of agricultural production.	Invasive ant impacts
Clarke, K.M., Fisher, B.L. & LeBuhn, G. 2008. *Urban Ecosyst* **11** pp. 317–334	The influence of urban park characteristics on ant (Hymenoptera, Formicidae) communities	• Urban forests of San Francisco reduced ant richness and abundance. • Invasive ant *Linepithema humile* had little or no impact to native ants of San Francisco.	Habitat change & invasive ant impacts
Fiedler, K. 2006. *Myrmecologische Nachrichten* **9** pp. 77–87.	Ant-associates of Palaearctic lycaenid butterfly larvae (Hymenoptera: Formicidae; Lepidoptera: Lycaenidae) – a review	• The number of Palaearctic lycaenid butterfly species correlates significantly with species richness of that ant genus attending it	Symbiosis
Andersen, A. 2008. *Journal of Biogeography* **24** pp. **399–539.**	Functional groups and patterns of organization in North American ant communities: a comparison with Australia	• Eight functional groups of ant communities of North America	Interactions of ants at food resources
Roth, D.S., Perfecto, I. & Rathcke, B. 1994. *Ecological Applications* **4** pp. 423–436.	The Effects of Management Systems on Ground-Foraging Ant Diversity in Costa Rica	Ant diversity at forest and abandoned cacao sites of Costa Rica was significantly more than from productive cacao and banana plantations	Impact of land use to ant Diversity.
Frouz, J, Holec,M. & Kalčík, J. 2003. *Pedobiologia* **47** pp. 205–212.	The effect of *Lasius niger* (Hymenoptera, Formicidae) ant nest on selected soil chemical properties	Most chemical elements and other parameters of the soil from the nests of ant *Lasius niger* were significantly higher than the surrounding soil.	Soil property Influence

Author, Year & Publication	Title	Summary of the study & findings	Ant Role Studies or Discovered
Dostál, P., Březnová, M., Kozlíčková, V., Herben, T. & Kovář,P. 2005. Pedobiologia **49** pp. 127–337.	Ant-induced soil modification and its effect on plant below-ground biomass	Ant *Lasius flavus* changes physical properties and distribute nutrients vertically from plant to access in their nests.	Soil property Influence
Gonthier, D.J., Ennis, K.K., Philpott, S.M., Vandermeer, J. & Perfecto, I. 2013. *BioControl* **58** pp. 815–820.	Ants defend coffee from berry borer colonization	Diverse ant species limit coffee pest beetle *Hypothenemus hampei* from colonizing coffee berries.	Agricultural biocontrol

Table 1.
List of some of publications related to the role of ants in the ecosystems.

when trees have grown big enough to accommodate the ant as the restoration age progresses. Relatively not much ant studies – on restoration – have been done investigate the conditions of ant species and the symbioses with other organisms. This further increase the untapped roles ants could be playing as indicators of restored ecosystems. (See **Table 1** for additional information about the roles of ants in ecosystems).

The relatively easy way to identify ants to genus, then to morphospecies level, further makes ants suitable to be added to the lists of indicators of restoration. The training one needs is to correctly identify the specimen using the external morphology of the specimen and properly following the identification keys provided by different ant taxonomic books and internet [50–54]. Which relatively does not take much time especially to someone with entomological background.

Identification of ant specimen also do not need relatively expansive equipment. A dissecting/stereo microscope with eye pieces with 10x magnification level and zooming ranges of 1x to 6.3x is enough to observe morphological features of the specimen when identifying them. Such microscopes satisfactory and clearly show ant body parts (such as antennal segments, wait segments, gaster segment etc.) that are commonly used to identify the specimen. They even easily allow one to increase or reduce the magnifications to accommodate the different parts of a specimen under observation, or change magnifications to accommodate specimen with different sizes. Following the identification keys and using a correct microscope are sufficient for one to correctly identify most specimen to genus or morphospecies levels [55].

The use of ant workers, instead of the males or the queens, as representatives when identifying their species, further makes ants more convenient to be used as indicators of restoration. Unlike their queens, that are rarely outside of the nests and very few relative to the workers, nor their male counterparts that are available temporarily, ant workers are most abundant and frequently outside the nests for easy sampling [37].

The flexibility of using different biodiversity indices from ant data also make ants suitable for indicating the restoration condition of the ecosystem. Biodiversity indices such as species richness and abundance are often used. Palladini et al. [56] used them to report the response of ant diversity to timber harvesting where they found that the high species richness and abundance of ants in sites with new harvest, relative to the sites with older harvests, indicate that ant community in those new harvested sites will take about hundred years after the disturbance to resemble the ones in mature forests. Carvalho & Vasconcelos [57] also used the species richness index to analyse the conditions of ants from fragmented forest relative to the continuous ones. The arranging ant data in functional groups of the sampled ants is also used to assess the ant communities from areas undergoing restoration. KING et al. [58] used it to assess the condition ant communities from the degraded forest sites where they found ant with opportunist functional group common at the disturbed sites relative to the reference sites, and almost all the tropical climate specialists and specialist predators were absent from the disturbed sites. Functional group was also used by Stephens and Wagner [59] to report that different functional groups dominate sites with different disturbances differently with the dominating groups at 'their' site suppressing or excluding other groups that less suited to the disturbance in that side. Ottonetti et al. [60] used both species richness and functional group plus diversity index to assess the response of ants to the rehabilitation done in their habitats where they found that richness and diversity index were not yet significantly different among the habitats but the functional groups were starting to change.

4. Sampling epigeal ants

There are four common methods used to sample the epigeal ants videlicet: hand collection, pitfall trapping, baiting and passive extraction[1]. Hand collection is one of the earliest methods used to collect ants where the specimen seen are directly picked up, from their nests or at the areas they are frequenting. Usually a pair of forceps or/and aspirators are used to pick up these specimen [54, 62–66] (see **Figure 1**).

Hand collection is efficient as it its data can produce ant species richness and more exclusive species [67]. It is often implemented in transects (of an appropriate lengths) in the study areas where ants seen are often collected at interval places of such transects [63, 65, 68, 69], for certain durations [69–71]. Sometimes hand collection is applied to the specific target area within the study site such as collecting specimen at the nest entrance or a specific microhabitat of interest [64, 67, 72, 73]. Cuautle et al. [54] used hand collection to sample the specimen of ants that were interacting with plant species at their sites to achieve the objective of their study investigates ant species that are consuming resources from the plants. Hand collection is often applied diurnally (usually from 7 h00 to 16 h00) when the collector can see ants and the weather conditions of the area are permitting and it assists in providing the information about ant species that are active at the specific time [54, 63, 64, 67].

Another common sampling technique for ground dwelling ants is pitfall trapping [74]. Pitfall trapping passively collects ants by capturing them after the ants fell into it when they are frequenting their habitats. Usually handle-less containers of different volumes are used as a pitfall traps [68, 69, 75–78] (see **Figure 2**). A set of pitfall traps are often spatially set at distance from each other [76, 79] for a certain durations ranging from 24 hours to two weeks [70, 80, 81].

It important to determine the efforts needed to set pitfall traps (such as the number of traps and the duration of trapping) that will produce the maximum ant information from the site/habitat/ecosystem at minimum resources before setting them. Gomes et al. [82]) found that only two days instead of 14 of pitfall trapping is needed to monitor ants of Ducke Reserve, Brazil which saved much needed time and funds. Andersen et al. [83] developed a further simplified pitfall trapping method that produced all the key findings of the intensive survey but with less than 10% of the intensive survey effort and made ant sampling achievable to anyone with limited ant training, time or resources.

---

[1] Extraction is more suitable for sampling areas with litter [61] and is hardly used by the authors.

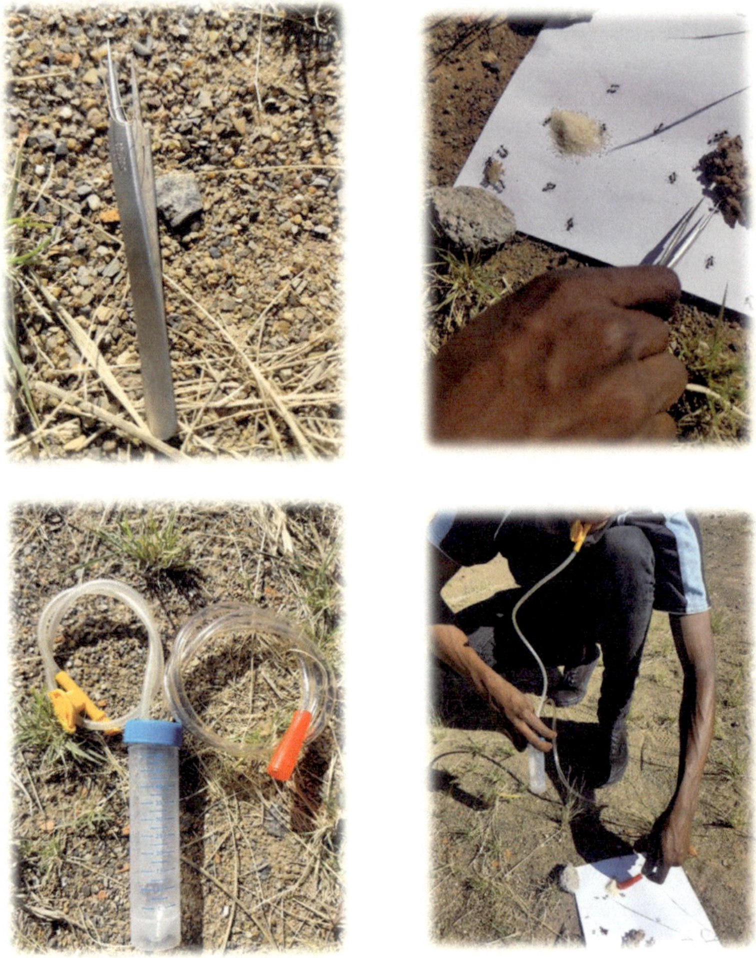

Figure 1.
Examples of a pair of forcepts (pictures above) and an aspirator (pictures above) and their collection demonstrations that were used to collect ants from Mokala National Park, South Africa.

Pitfall traps are buried in a ground, with their rims at level to the ground surface (see **Figure 3**) where a half to two-thirds of their capacity is often filled with a solution of water- propylene glycol (car antifreeze/coolant) to impound, kill and preserve the ants [83, 84]. Sometimes a soapy water is also used [68, 69]. Pitfall traps catch ants continuously on their own for a relative long time, at different times, without human presence. It therefore provide a wider information about ant diversity of the habitat as a whole and is one of the effective sampling methods to estimate ant species richness in a concerned study site [85].

Another most common technique to sample ants is baiting. Similar to hand collection, baiting is also often implemented in transects (of an appropriate lengths) in the study areas where ants seen. Baiting stations are often placed at interval places of such transects where foods such as tuna, sardines, biscuits, water-sugar or water-honey solutions are placed on material such as petri dishes, vials or pieces of paper [86–89] (see **Figure 4**). Bating is left operating for certain durations [90–92]. Sometimes baiting can be placed at a specific target area depending of the objective of the study. For instance Tonhasca [93] placed baiting on the trail of ant *Atta sexdens rubropilosa* as they were investigating the behaviour of *Neodohrniphora declinata*, a parasitic fly of the ant, roosts along ants' foraging trails. Baiting is often

Figure 2.
Containers that are commonly used as pitfall traps to sample ground dwelling ants in South Africa.

Figure 3.
A pitfall trap set in the ground to catch ants from a site that was heavily degraded by livestock grazing in Tankwa Karoo National Park, South Africa.

applied diurnally (usually from 8 h20 to 18 h00) when the collector can see ants and their interactions [94] and it can be done when weather conditions of the area are permitting [95].

Baiting enable a researcher to record the interactions amongst ant species/colonies and provide the mechanisms the different species employ to access resources such as food [96, 97]. Peral et al. [98] used it to investigate if the dominating ant species could be restrictively influencing the traits of the subordinate ones in

Figure 4.
Example of baits (viz. tuna on left and sugar on right) on a white paper sheet (aiding to easy see ants), with some ant species consuming them, that were used to sample ants from Mountain Zebra National, South Africa.

habitat where they discovered that the presence of dominant ant species limited the abundance and occurrence of subordinates. It can also be used to investigate the interactions amongst individuals of the same ant species from different colonies. Garnas et al. [99] also used baiting to study the magnitude of intercolony aggression amongst the invasive *Myrmica rubra* ant colonies in Maine, USA, where they discovered that *M. rubra* foragers that originate from the same colony, that has however fragmented, still have the high levels of intraspecific tolerance and has intercolony aggression to foragers from neighbouring colonies. Baiting also assists in studying or identifying the mechanisms some ant species employ to be competitive/dominant to other species. Holway [100] used it to learn that the invasive *Linepithema humile* in California, uses the faster resource discovery and faster member recruitments than the native species as one of the mechanisms it dominate the ecosystem.

Baiting also assist in studying the activities of different ant species at different climatic conditions, and the influence such abiotic factors have to them [101–103]. For instance, Cerdá et al. [104] used it to learn that the foraging activities of the sub-dominant ant species of the open grassland habitat of the southeastern Spain is influenced more by the temperature of their habitat than the competition with their dominant counterparts.

For all these sampling methods, a pair of forceps and an aspirator (**Figure 1**), are used to transfer the specimen into containers such as vials that mostly contain a 75%, or more, alcohol solution to preserve the specimen [105, 106]. A magnifying glass (**Figure 5**), can also be used in the field to enlarge the ant specimen, and to make the observations of their interactions clearer.

In most cases the specimen are transported to the laboratories where they are identified to the genus and morphospecies levels following the appropriate taxonomic books and internet websites [107].

The efficiency of these methods in producing the highest biodiversity information of ants can sometimes differ. It is therefore prudent to assess the most efficient one for the habitats/ecosystem one is going to sample. Although some sampling methods are more efficient in getting more species, applying the combination of different methods is often the most effective way of sampling ants. King and Porter [108] reported that the combination of baiting, pitfall trapping, extraction and hand collecting was more efficient in sampling more ant species than using only one method as these methods complement each other on species that were not collected

Figure 5.
Example of a magnifying glass, and its collection demonstration that were used to sample ant specimen from Mokala National Park, South Africa.

by other methods. Antoniazzi et al. [67] also reported that although hand collection sampled more ant species in the Mexican rainforest than baiting, the combination of these methods yielded more ant species than just one method. These methods can be implemented simultaneously in the habitat undergoing restoration to accommodate wider variables - and get bigger picture - related to ant conditions. Pitfall trapping, which is a passive sampling method, will catches ants continuously on its own for a relative long time, at different times – diurnal to nocturnal and provide ant specimen for diversity information even when human is not around. It can however can be complemented by baiting and hand collection that will record the interactions amongst ant species/colonies in the habitat and provide the mechanisms the different species employ to access resources such as food [96, 97].

The efficiency of sampling methods can also be assessed by using estimators (such as incidence-based coverage estimator (ICE), Jackknife estimator (Jack2) etc.). Lopes and Vasconcelos [68] and de Souza et al. [85] used estimators to evaluate the efficiency of the three methods (viz. sardine baiting, pitfall trapping and extractor) at the savanna and forest habitats of Brazil.

5. Analysing the sampled ant data

There are many calculations that can be used to analyse and compare the collected ant data. One of the first analyses to do is to determine if the species of the sampled habitat/site have been adequately sampled, or determine the sampling efforts needed to represent the adequate ant species of the habitat/site/ecosystem. This is often achieved by calculating the estimated species that should be at the habitats/ecosystems relative to the ones sampled from the data collect, and then determines if the number of the sampled species is sufficient relative to the expected ones [63, 67, 68, 71, 78, 109–111]. For instance, Urrutia-Escobar and Armbrecht [112] used the averages of the three estimator (ICE, Chao2) and Jack1) to determine that they have sampled about 85% of the expected ant species of their study sites while Cerdá et al. [113] used Chao1 estimator to decide that the number of ant species they have sampled was similar to the expected ones for the habitat. ICE, Jack2 and Michaelis–Menten richness estimator (MMMean) was used by Calcaterra et al. [77] to find that ant species they sampled were below the expected of the habitats they studied.

The expected ant species and the amount of sampling time required for a site/habitat/ecosystem can be calculated by using species accumulation curves as they can identify the size of the sampling site and the sampling period needed to get the maximum ant species in a habitat/ecosystem. Wang et al. [61] used them to assess that to capture the maximum ant species, using pitfall traps at George Washington and Monongahela National Forests of USA, one needs to sample eight plots for eight weeks (see **Figure 6**).

One of the basic and common indices that are used to analyse the diversity of ants in sampled habitats/sites is species richness. The ant species richness is the total number of ant species that has been sampled from the study sites [114]. Finding the species richness can be done by just counting these species from the sampled data of the concerned site/habitat/ecosystem or period. Species richness can be used to assess the conditions of the habitats under study like King et al. [58] and Graham et al. [84] did to evaluate the impact of habitat disturbances and found low species richness on sites of the disturbed areas relative to their undisturbed counterparts. Porter and Savignano [115] used it to find that the invasive *Solenopsis invicta* ant has drastically reduced the native ant community of Texas, USA.

Another basic index that is used to analyse the diversity of ants at a particular habitat/ecosystem/site is abundance. Abundance is the total number of ant individuals that have been sampled [116], and can be found by just counting the number of the ant individuals sampled. Just like any comparative analysis indices, abundance can be used to evaluate and compare the ecological conditions amongst sites/habitats/ecosystems or of interest. Morrison [117] used it to assess the short and long term impact of the ant *Solenopsis invicta* invasion to the native ant species of Texas, USA, and discovered that its impact is more severe at the beginning of the invasion and subsides as time goes on. Rivas-Arancibia et al. [118] also used the abundance from ant data collected from in Puebla, Mexico to learn that the most abundant ant species, at both the disturbed and undisturbed sites, are more active in mornings than in the evenings.

Abundance can also be used to determine the ant species that are numerically dominating the habitats. Sarty et al. [119] and Parr [120] both used it to investigate

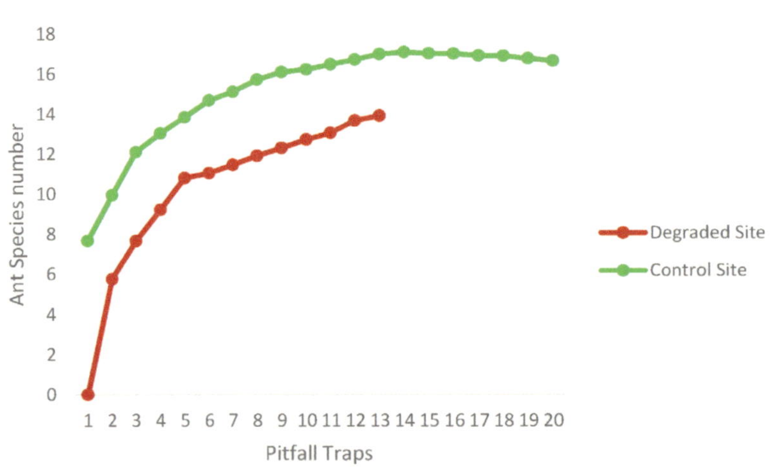

Figure 6.
An example of species accumulative curves that shows ant species from a site undergoing restoration relative to the control one.

the dominating ant species at different habitats of Tokelau, New Zealand and Kruger National Park, South Africa, respectively. The information about the abundance also assists to investigate the association of different ant species to different habitats in at different ecosystems. Calcaterra et al. [77] used it to learn that the cosmopolitan invasive ant, *Solenopsis invicta*, dominates both the natural and the modified habitats in its native range of Corrientes, Argentina. Calcaterra et al. [121] also used it to study the competitive mechanisms of the other cosmopolitan invasive ant *Linepithema humile*, in its native range, and learned that it recruits its members quickly to the food resources.

The abundance can also be used to identify ant species that could be the indicators of the conditions of their habitats. Herrera-Rangel et al. [81] learned that ant *Gnamptogenys bisulca* could be an indicator species of a healthy forests (that have thick leaf litter profiles and high relative humidity values) of Central Mountain Range, in Colombia more abundant in such habitats.

The calculations of both species richness and abundance of ants can be done by many programmes available of different platforms. Abundance is one of the basic components of other analyses that also investigate the biodiversity of ants of a particular habitat or ecosystem such as Indicator Value Analysis index (IndVal) that identifies indicator species of the habitat/ecosystem under study [122, 123]. With the inclusion of the abundance in IndVal, Sanabria et al. [124] identified 14 ant species that can indicate the ecosystem service condition of the soils from five different land uses.

6. Basic gradual progress to use ant data for rehabilitation processes

1. Choose sampling sites.

    a. Choose a rehabilitated site. This is a place/s where the rehabilitation is taking place from degradation (see **Figure 7**).

Figure 7.
An example of places where rehabilitation site (forefront of the red line in the picture) and the reference site (behind the red line in the picture) were set to sample ants to assess the ant diversity conditions of the rehabilitated site relative to their natural conditions [125].

i. Choose a reference site (in cases where available) (see **Figure** 7). This is a place/s that has not experienced the concerned degradation. Ant diversity from the rehabilitation site/s will be compared to the ant diversity from the reference site to determine how different ant diversity of rehabilitation site from their natural diversity.

b. If possible, replicate sampling sites.

2. Select an appropriate ant sampling method and durations related to your objectives (see section four).

3. If possible do a pilot sampling for each side to determine sampling efforts that will be needed in each site [61].

4. Execute the sampling methods accordingly.

5. Ensure that step 1 to 4 are carefully considered as they are the most expensive (in both monetary- and manpower-wise) to rectify in case they were miscalculated.

6. Prepare specimen for identification – under the microscope.

a. For pitfall trapping remove ant specimen from none-ant at each trap.

i. Use basic characteristics of ants to distinguish them from none-ants of each trap (viz. a pair of antenna (consisting funiculus and a scape) that is usually jack-knifed and a petiole/wait (consisting of either a single or two segments) (see **Figure 8**).

b. Group the specimen according to the similarity of their morphology such as their sizes and colours.

7. Identify specimen to the taxonomical rank appropriate to your objectives.

8. Determine if the species of the sampled site have been adequately sampled by using appropriate species accumulation curves.

9. Determine the diversity of each side and interpret the results.

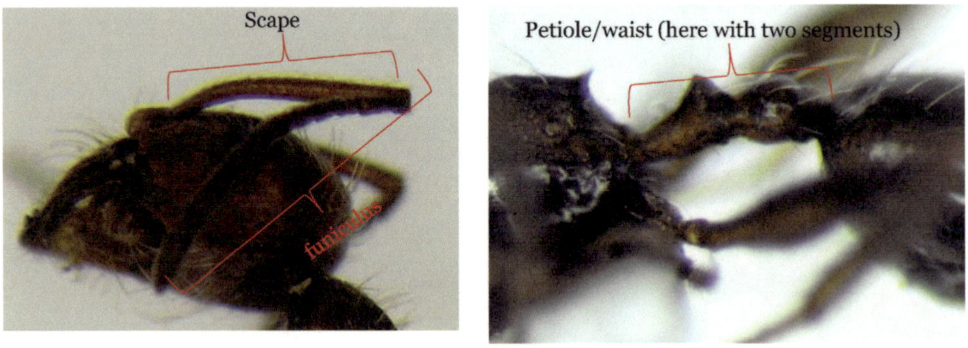

Figure 8.
A close view of antennal and waist parts of ants use distinguish ants from non-ant insects.

7. Ant biodiversity indices that can be used indicators of restoration

Biodiversity indices can show species that are associated with or indicate the conditions of their habitats/sites – therefore be the indicators of those habitats/sites. For example abundance reduction of ants can indicate the condition of their habitats as it did on land transformed into urbanisation in in Indiana, U.S.A [126]. Ant abundance also indicates the condition of the degraded habitats relative to the un-degraded ones like it was significantly lower on habitats turned into oil plantation relative to the intact forest in Sabah, Malaysia [127]. Ant species also can indicate the conditions of their habitats like *Pachycondyla impressa* indicated the influence of shading to its habitat as it was found on shaded coffee plantation that accommodated its natural habitat requirements of the forest floor, than at unshaded ones [112].

The other ant indicator of the habitats/sites can be the ant community composition as it can also indicate the conditions of the habitats/sites. For instance, the ant community compositions of habitats undergoing forest restoration in Colombia were still having different ant community compositions to the pristine forests [81].

The presence of ant species can also be used as indicator of the state of the habitats. The presence of some species such as *Crematogaster cerasi* and *Prenolepis impairs,* can indicate that status of the degradation at the site/habitat/ecosystem undergoing restoration as such species disappeared at the presence of degradation like they did after the residential development in Indiana, U.S.A [126]. Some such as tramp ant species (which are alien and invasive species) can also indicate the conditions of habitats/sites that have been degraded as they did at habitats that have been turned into rubber and oil plam plantations relative to forest in Sumatra, Indonesia [128].

Author details

Hendrik Sithole[1,2]* and Nolubabalo Tantsi[3]

1 Scientific Services, South African National Parks, Kimberley, South Africa

2 School of Animal, Plant and Environmental Sciences, University of the Witwatersrand, Johannesburg, South Africa

3 Department of Agriculture and Rural Development, Provincial Government of South Africa, Johannesburg, South Africa

*Address all correspondence to: hendrik.sithole@sanparks.org

References

[1] Folgarait PJ. Ant biodiversity and its relationships to ecosystem functioning: A review. Biodiversity and Conservation. 1998; 7: 1221-1244.

[2] Bamba D, Bessiere, JM, Marion C, Pelisser Y, Fouraste I. Essential oil Eupatorium odoratum. Planta Medica. 1993; 59: 184-185.

[3] Lonsdale WM. Global patterns of plants invasions and the concept if invisibility. Ecology. 1999; 80: 1522-1536.

[4] Dregne HE. Desertification of arid lands. Harwood Academic Publishers. 1983; Chur, Switzerland.

[5] Rocha-Nicoleite E, Overbeck GE, Muller SC. Degradation by coal mining should be priority in restoration planning. Perspectives in Ecology and Conservation. 2017; https://doi.org/10.1016/j.pecon.2017.05.006.

[6] Tilman DH, Knops J, Wedin D, Reich P, Ritchie M, Siemann E. The influence of functional diversity and composition on ecosystem processes. Science. 1997:277:1300-1302.

[7] Evans RD, Rimer R, Sperry L, Belnap J. Exotic plant invasion alters nitrogen dynamics in an arid grassland. Ecological Applications. 2001; 11:1301-1310.

[8] SER. The SER International Primer on Ecological Restoration. Version 2. Society for Ecological Restoration International Science and Policy Working Group. 2004 https://cdn.ymaws.com/www.ser.org/resource/resmgr/custompages/publications/SER_Primer/ser_primer.pdf.

[9] Holl KD, Aide TM. When and where to actively restore ecosystems? Forest Ecology and Management. 2011:261:1558-1563.

[10] Keenleyside KA, Dudley N, Cairns S, Hall CM, Stolton S. Ecological restoration for protected areas: principles, guidelines and best practices. 2012. IUCN, Gland, Switzerland.

[11] UNEP-WCMC, IUCN. Protected Planet: The World Database on Protected Areas (WDPA) [Online], September 2020, Cambridge, UK: UNEP-WCMC and IUCN. Available at: www.protectedplanet.net.

[12] Clark NE, Boakes EH, McGowan PJK, Mace GM, Fuller RA. Protected Areas in South Asia Have Not Prevented Habitat Loss: A Study Using Historical Models of Land-Use Change. PLoS ONE. 2013;8: e65298. https://doi.org/10.1371/journal.pone.0065298.

[13] Jones KR, Venter O, Fuller RA, Allan JR, Maxwell SL, Negret PJ, Watson JEM. One-third of global protected land is under intense human pressure. Science. 2018;360:788-791. DOI: 10.1126/science.aap9565.

[14] Dudley N. (ed.). Guidelines for Applying Protected Area Management Categories. IUCN, Gland, Switzerland. https://portals.iucn.org/library/sites/library/files/documents/PAG-021.pdf. 2008.

[15] Convention on Biological Diversity (Secretariat). The strategic plan for biodiversity 2011-2020 and the Aichi biodiversity targets. Document UNEP/CBD/COP/DEC/X/2. Secretariat of the Convention on Biological Diversity, Nagoya, Japan. https://www.cbd.int/sp/targets/. 2010.

[16] ten Brink, P (ed). The Economics of Ecosystems and Biodiversity in National and International Policy Making. An output of TEEB: The Economics of Ecosystems and Biodiversity. 2011. Earthscan, London.

[17] UNEP-WCMC. Ten years to boost ecosystem restoration for people and

planet https://www.unep-wcmc.org/news/10-years-to-boost-ecosystem-restoration-for-people-and-planet. 2020.

[18] Convention on Biological Diversity (Secretariat). Considerations on ecosystem restoration for the post-2020 global biodiversity framework, including on a possible successor to Aichi biodiversity target 15. CBD/POST2020/WS/2019/11/3. 2020. Secretariat of the Convention on Biological Diversity, Rio de Janeiro, Brazil https://www.cbd.int/doc/c/fcd6/bfba/38ebc826221543e322173507/post2020-ws-2019-11-03-en.pdf.

[19] Ferrier S. Prioritizing where to restore Earth's ecosystems. Nature. 2020; 586:680-681.

[20] Gichuki L, Brouwer R, Davies J, Vidal A, Kuzee M, Magero C, Walter S, Lara P, Oragbade C, Gilbey B. Reviving Land and Restoring Landscapes: Policy Convergence between Forest Landscape Restoration and Land Degradation Neutrality; IUCN Publication: Gland, Switzerland, 2019.

[21] United Nations Sustainable Development, About the Sustainable Development Goals. https://unstats.un.org/sdgs/report/2020/The-Sustainable-Development-Goals-Report-2020.pdf. 2020.

[22] Dave R, Saint-Laurent C, Murray L, Antunes Daldegan G, Brouwer R, de Mattos Scaramuzza CA, Raes L, Simonit S, Catapan M, García Contreras G, Ndoli A. 2018. Second Bonn Challenge progress report. Application of the Barometer in 2018.

[23] UNEP. *New UN Decade on Ecosystem Restoration offers Unparalleled Opportunity for Job Creation, Food Security and Addressing Climate Change*. 2019. https://www.unenvironment.org/news-and-stories/press-release/new-un-decade-ecosystem-restoration-offers-unparalleled-opportunity.

[24] Muñoz-Rojas, M. Soil quality indicators: critical tools in ecosystem restoration. Current Opinion in Environmental Science & Health. 2018;5:47-52.

[25] Doren RF, Trexler JC, Gottlieb A, Harwell MC. Ecological indicators for system-wide assessment of the greater everglades ecosystem restoration program. Ecological Indicators. 2009;9:S2-S16.

[26] Liddicoat C, Weinstein P, Bissett A, Gellie NJC, Mill, JG, Waycott M, Breed MF. Can bacterial indicators of a grassy woodland restoration inform ecosystem assessment and microbiota-mediated human health? Environment International. 2019; 129:105-117.

[27] Spain AH, Hinz D, Tibbett M. Colonisation of rehabilitated lands by termites (Dictyoptera), Rio Tinto Alcan Gove bauxite mine, Northern Territory, Australia, in Proceedings Fifth International Conference on Mine Closure (Mine Closure 2010), A.B. Fourie, M. Tibbett and J. Wiertz (eds), Viña del Mar, Chile, Australian Centre for Geomechanics, Perth. 2010; 437-448.

[28] Feest A, van Swaay C, van Hinsberg A. Nitrogen deposition and the reduction of butterfly biodiversity quality in the Netherlands. Ecological Indicators. 2014; 39:115-119.

[29] McKinley VL, Peacock AD, White DC. Microbial community PLFA and PHB responses to ecosystem restoration in tallgrass prairie soils. Soil Biology & Biochemistry. 2005;37:1946-1958.

[30] McGoech MA. The selection, Testing, and Application of Terrestrial insects as Bioindicators. Biological Reviews. 1998;73:181-201.

[31] Majer JD. Ants: Bio-indicators of minesite rehabilitation, land use and

land conservation. Environmental Management. 1983;7:375-383.

[32] Luque GM, Reyes-López, J, Fernández-Haeger J. Recovery of Ground Ant (Hymenoptera: Formicidae) communities six years after a major environmental disaster. Entomological Society of America. 2007;36:337-347.

[33] Gutierrez-Rapalino BP, Domínguez-Haydar Y. Contribution of Pheidole fallax and Ectatomma ruidum (Hymenoptera: Formicidae) to the dispersion and germination of seeds in rehabilitated areas of the Cerrejón coal mine, Colombia. Revista de Biología Tropical. 2016;65:575-587.

[34] Domínguez-Haydar Y, Gutierrez-Rapalino B, Jiménez JJ. *Density and Spatial Distribution of Nests of Ectatomma ruidum and Pheidole fallax (Hymenoptera: Formicidae), as Response to the Recovery of Coal Mine Areas.* Sociobiology. 2018; *65:415-421.*

[35] Zettler JA, Taylor MD, Allen CR, Spira TP. Consequences of Forest Clear-Cuts for Native and Nonindigenous Ants (Hymenoptera: Formicidae). Annals of the Entomological Society of America. 2004;97:513-518.

[36] Andersen AN, Sparling GP. Ants as indicators of restoration success: relationship with soil microbial biomass in the Australian seasonal tropics. Restoration Ecology. 1997; 5:109-114.

[37] Hölldobler B, Wilson EO. *The ants*. Springer-Verlag Berlin Heidelberg. 1990.

[38] Wang, D., McSweeney, K., Lowery, B. & Norman, J.M. 1995. Nest structure of ant Lasius neoniger Emery and its implications to soil modification. Geoderma **66** pp. 259-272.

[39] Moreira, A.A., Forti, L.C., Boaretto, M.A.C., Andrade, A.P.P., Lopes J.F.S. & Ramos, V.M. 2004. External and internal structure of *Atta bisphaerica* Forel (Hymenoptera: Formicidae) nests. Journal of Applied Entomology 128 pp. 204-211.

[40] Hasegawa, E. 1993. Nest defense and early production of the major workers in the dimorphic ant Colobopsis nipponicus (Wheeler) (Hymenoptera: Formicidae). Behavioral Ecology and Sociobiology 33:73-77.

[41] Bestelmeyer A, Wiens J. The effects of land use on the structure of ground-foraging ant communities in the Argentine Chaco. Ecological Applications. 1996;6:1225-1240.

[42] Liefke C, Dorow W, Hölldoble B, *et al.* Nesting and food resources of syntopic species of the ant genus *Polyrhachis* (Hymenoptera, Formicidae) in West-Malaysia. Insectes sociaux. 1998;45:411-425.

[43] Díaz, M. Spatial patterns of granivorous ant nest abundance and nest site selection in agricultural landscape of Central Spain. 1991.

[44] Sorvari J, Hakkarainen H. Wood ants are wood ants: deforestation causes population declines in the polydomous wood ant Formica aquilonia. Ecological Entomology. 2007;32:707-711.

[45] Maurizi E, Fattorini S, Moore W, Di Giulio A. Behavior of Paussus favieri (Coleoptera, Carabidae, Paussini): A Myrmecophilous Beetle Associated with *Pheidole pallidula* (Hymenoptera, Formicidae). Psyche. 2012. Article ID 940315, 9 pages; doi:10.1155/2012/940315.

[46] Pringle, E.G., Dirzo, R. & Gordon, D.M. 2011. Indirect benefits of symbiotic coccoids for an ant-defended myrmecophytic tree. Ecology, 92 pp. 37-46.

[47] Currie CR. A community of ants, fungi, and bacteria: a multilateral approach to studying symbiosis. Annual Review Microbiology. 2001;55:357-380.

[48] Pierce NE, Elgar MA. Behavioral Ecology and Sociobiology. 1985;16: 209-222.

[49] Hazra A. *Recruitment Dynamics of Vachellia cornigera and its ant occupants in restored tropical forest*. University of Illinois at Chicago. Thesis. 2018; https://hdl.handle.net/10027/22668.

[50] Petersen, CE, Zwolfer K, Frandkin J. Ant fauna of reconstructed tallgrass prairie in Northeastern Illinois. Transections of the Illinois State Academy of Science. 1998; 91:85-90.

[51] Abbott KL, Sarty M, Lester PJ. The ants of Tokelau. – New Zealand. Journal of Zoology. 2006;33:157-164.

[52] Oliveira MA, Della Lucia TMC, Marinho CGS, Delabie JHC, Morato, EF. Ant diversity in an area of the Amazon forest in Acre, Brazil. Sociobiology. 2009;54:243-267.

[53] Kajzer-Bonk J, Szpiłyk D, Woyciechowski M. Invasive goldenrods affect abundance and diversity of grassland ant communities (Hymenoptera: Formicidae). Journal of Insect Conversion. 2016;20:99-105. https://doi.org/10.1007/s10841-016-9843-4.

[54] Cuautle M, Castillo-Guevara C, Juárez-Juárez B, Pérez-Toledo G. Ants (Hymenoptera: Formicidae) in a temperate ecosystem from La Malinche National Park, Mexico. Florida Entomologist. 2020;103:321-328.

[55] Le Breton J, Jourdan H, Chazeau J, Orivel J, Dejean A. Niche opportunity and ant invasion: the case of Wasmannia auropunctata in a New Caledonian rain forest. Journal of Tropical Ecology. 2005;21:93-98.

[56] Palladini JD, Jones MG, Sanders NJ, Jules ES. The recovery of ant communities in regenerating temperate conifer forests. Forest Ecology and Management. 2007;242:619-624.

[57] Carvalho KS, Vasconcelos HL. Forest fragmentation in central Amazonia and its effects on litter-dwelling ants. Biological Conservation. 1999;91:151-157.

[58] King JR, Andersen AN, Cutter AD. Ants as bioindicators of habitat disturbance: validation of the functional group model for Australia's humid tropics. Biodiversity and Conservation. 1998;7:1627-1638.

[59] Stephens SS, Wagner MR. Using Ground Foraging Ant (Hymenoptera: Formicidae) Functional Groups as Bioindicators of Forest Health in Northern Arizona Ponderosa Pine Forests. Environmental Entomology. 2006;35:937-949.

[60] Ottonetti L, Tucci L, Santini G. Recolonization patterns of ants in a rehabilitated lignite mine in central Italy: potential for the use of Mediterranean ants as indicators of restoration processes. Restoration Ecology. 2006;14:60-66.

[61] Wang, C., Strazanac, J. & Butler, L. 2001. A comparison of pitfall traps with bait traps for studying leaf litter ant communities. Journal of Economic Entomology 94 pp. 761-765.

[62] Agosti D, Majer JD, Alonso LE, Schultz TR. Ants: Standard Methods for Measuring and Monitoring Biodiversity. Smithsonian Institution Press, Washington, D.C. 2000.

[63] Rizali A, Bos MM, Buchori D, Yamane S, Hans C, Schulze J. Ants in tropical urban habitats: the myrmecofauna in a densely populated area of Bogor, West Java, Indonesia. Hayati Journal of Biosciences. 2008; 15:77-84.

[64] Ouellette GD, Drummond FA, Choate B, Groden E. Ant diversity and distribution in Acadia National Park, Maine. Environmental Entomology. 2010;39:1447-1456. DOI:10.1603/EN09306.

[65] Watanasit S, Nhu-eard T. Diversity of ants (Hymenoptera: Formicidae) in two rubber plantations in Songkhla Province, Southern Thailand. Songklanakarin Journal of Science and Technology. 2011;33:151-161.

[66] Sosa-Calvo J, Jesovnik A, Okonski E, Schultz TR. Locating, collecting, and maintaining colonies of fungus-farming ants (Hymenoptera: Myrmicinae: Attini). Sociobiology. 2015;62:300-320.

[67] Antoniazzi R, Ahuatzin D, Pelayo-Martinéz J, Ortiz-Lozada L, Leponce M, Dáttilo W. On the effectiveness of hand collection to complement baits when studying ant vertical stratification in tropical rainforests. Sociobiology. 2020;67:213-222. DOI: 10.13102/sociobiology.v67i2.4909.

[68] Lopes CT, Vasconcelos HL. Evaluation of three methods for sampling ground-dwelling ants in the Brazilian cerrado. Neotropical Entomology. 2008;37:399-405.

[69] Dias RKS, Perera KAM. Worker ant community observed by repeated sampling and information on endemic Aneuretus simoni Emery in the Gilimale Forest Reserve in Sri Lanka. Asian Myrmecology. 2011;4:69-78.

[70] Sukimin S, Mohamed M, Aris H. Ant diversity of Maliau Basin Conservation area, Sabah, Malaysia. Journal of Tropical Biology and Conservation. 2010;6:89-101.

[71] Bernadou A, Céréghino R, Barcet H, Combe M, Espadaler X, Fourcassié V. Physical and land-cover variables influence ant functional groups and species diversity along elevational gradients. Landscape Ecology. 2013; 28: 1387-1400.

[72] Burwell CJ, Nakamura A. Distribution of ant species along an altitudinal transect in continuous rainforest in subtropical Queensland, Australia. Memoirs of the Queensland Museum. 2011;55:391-411.

[73] Fernandez-Marin H, Bruner G, Gomez EB, Nash DR, Boomsma JJ, Wcislo WT. Dynamic disease management in Trachymyrmex fungus-growing ants (Attini: Formicidae). American Naturalist. 2013;181:571-582.

[74] Borgelt A, New T. Pitfall trapping for ants (Hymenoptera, Formicidae) in mesic Australia: the influence of trap diameter. Journal of Insect Conservation, 2005;9:219-221. https://doi.org/10.1007/s10841-005-5172-8.

[75] Peck SL, McQuaid B, Campbell CL, Using ant species (Hymenoptera: Formicidae) as a biological indicator of agroecosystem condition. Environmental Entomology. 1998; 27:1102-1110.

[76] Wang C, Strazanac J, Butler L. Abundance, diversity, and activity of ants (Hymenoptera: Formicidae) in oak-dominated mixed Appalachian forests treated with microbial pesticides, Environmental Entomology. 2000;29:579-586.

[77] Calcaterra LA, Cuezzo F, Cabrera SM, Briano JA. Ground ant diversity (Hymenoptera: Formicidae) in the Ibera Nature Reserve, the largest wetland of Argentina. Conservation Biology and Biodiversity. 2010;03:71-83.

[78] Tăusan I, Dauber J, Trică MR, Markó B. Succession in ant communities (Hymenoptera: Formicidae) in deciduous forest clear-cuts – an Eastern European case study. European Journal of Entomology. 2017;114:92-100.

[79] Zhao ZH, Shi PJ, Hui C, Ouyang F, Ge F, Li B.L. Solving the pitfalls of pitfall trapping: a two-circle method for density estimation of ground-dwelling arthropods. Methods in Ecology and Evolution. 2013;4:865-871.

[80] Kwon T, Lee CM. Sung JH. Diversity decrease of ant (Formicidae, Hymenoptera) after a forest disturbance: different responses among functional guilds. Zoological Studies.2014;53: https://doi.org/10.1186/s40555-014-0037-z

[81] Herrera-Rangel J, Jiménez-Carmona E, Armbrecht I. Monitoring the diversity of hunting ants (Hymenoptera: Formicidae) on a fragmented and restored Andean landscape. Environmental Entomology. 2015;44:1287-1298. https://doi.org/10.1093/ee/nvv103.

[82] Gomes CB, Souza JLP, Franklin E. A comparison between time of exposure, number of pitfall traps and the sampling cost to capture ground-dwelling poneromorph ants (Hymenoptera: Formicidae). Sociobiology. 2018;65: 138-148.

[83] Andersen A. Ants as indicators of ecosystem restoration following mining. A functional group approach. In: Conservation Outside Nature Reserves (eds P. Hale and D. Lamb) Centre for Conservation Biology, The University of Queensland, Brisbane. 1997;319-325.

[84] Graham JH, Hughie HH, Jones S, Wrinn K, Krzysik A., Duda JJ, Freeman DC, Emlen JM, Zak JC, Kovacic DA, Chamberlin-Graham C, Balbach H. Habitat disturbance and the diversity and abundance of ants (Formicidae) in the Southeastern FallLine Sandhills. Journal of Insect Science. 2004;4:1-15.

[85] de Souza JLP, Baccaro FB, Landeiro VL, Elizabeth F, Magnusson WE. Trade-offs between complementarity and redundancy in the use of different sampling techniques for ground-dwelling ant assemblages. Applied Soil Ecology. 2012; 56:63-73.

[86] Marques GDV, K. Del Claro. The ant fauna in a Cerrado area: The influence of vegetation structure and seasonality (Hymenoptera: Formicidae). Sociobiology. 2006;47:235-252.

[87] Fayle TM, Bakker L, Cheah C, Ching TM, Davey A, Dem F, Earl A, Huaimei Y, Hyland S, Johansson B, Ligtermoet E, Lim R, Lin LK, Luangyoth P, Martins BH, Palmeirim AF, Paninhuan S, Rojas SK, Sam L, Sam PTT, Susanto D, Wahyudi A, Walsh J, Weigl S, Craze PG, Jehle R, Metcalfe D, Trevelyan R. A positive relationship between ant biodiversity (Hymenoptera: Formicidae) and rate of scavenger-mediated nutrient redistribution along a disturbance gradient in a southeast Asian rain forest. Myrmecological News. 2010;14:5-12.

[88] Taxonomic and functional diversity of ants (hymenoptera: formicidae) in an upper hill Dipterocarp forest in peninsular Malaysia. Mustafa NA, Salim HMW, Fletcher C, Kassim AR, Potts MD. The Raffles Bulletin of Zoology .2011;59:181-194.

[89] Choate BA, Drummond FA. Ant diversity and distribution (Hymenoptera: Formicidae) throughout Maine lowbush blueberry fields in Hancock and Washington counties. Journal of Environmental Entomology. 2012;41: 222-232.

[90] Yanoviak S.P., Fisher B.L. & Alonso A. 2007. Arboreal ant diversity (Hymenoptera: Formicidae) in a central African forest. African Journal of Ecology 46 pp. 60-66. doi:10.1111/j.1365-2028.2007.00810.x.

[91] Asfiya, W.A.R.A., Lach, L., Majer, J.D., Heterick, B.R.I.A.N. & Didham, R.K. 2015. Intensive agroforestry practices negatively affect ant (Hymenoptera: Formicidae) diversity and composition in southeast Sulawesi, Indonesia. Asian Myrmecology 7 pp. 87-104.

[92] Masse, P.S.M., Tindo, M., Djiéto-Lordon, C., Mony, R. & Kenne, M. 2019. Diversity of ant assemblages

(Hymenoptera: Formicidae) in an urban environment in Cameroon during and after colonization of the area by Wasmannia auropunctata. European Journal of Entomology 116 pp. 461-467.

[93] Tonhasca, A.J. 1996. Interactions between a parasitic fly, Neodohrniphora declinata (Diptera: Phoridae), and its host, the leaf-cutting ant Atta sexdens rubropilosa (Hymenoptera: Formicidae). Ecotropica 2 pp. 157-164.

[94] Eguchi, K. & Yamane, S. 2003. Species diversity of ants (Hymenoptera, Formicidae) in a lowland rainforest, Northwestern Borneo. New Entomologist 52 pp. 49-59.

[95] Czechowski, W., Markó, B., Radchenko, A. & Ślipiński, P. 2013. Long-term partitioning of space between two territorial species of ants (Hymenoptera: Formicidae) and their effect on subordinate species. European Journal of Entomology 110 pp. 327-337.

[96] Perfecto I. 1994. Foraging behavior as a determinant of asymmetric competitive interaction between two ant species in a tropical agroecosystem. Oecologia 98 pp. 184-192.

[97] Human, K.G. & Gordon. D.M. 1996. Exploitation and interference competition between the invasive I.inepithema humile, Argentine ant, and native ant species. Oecologia 105 pp. 405-412.

[98] Peral, G.T., Markó, B., Babik, H., Tăuşan, I., Maák, I., Pálfi, Z., Ślipiński, P., Czekes, Z. & Czechowski, W. 2016. Differential Impact of Two Dominant Formica Ant Species (Hymenoptera, Formicidae) on Subordinates in Temperate Europe. Journal of Hymenoptera Research 50 pp. 97-116.

[99] Garnas, J.R., Drummond, F.A. & Groden, E. 2007. Intercolony Aggression Within and Among Local Populations of the Invasive Ant, Myrmica rubra (Hymenoptera: Formicidae), in Coastal Maine. Environmental Entomology 36 pp. 105-113.

[100] Holway D.A. Competitive mechanisms underlying the displacement of native ants by the invasive argentine ant. Ecology. 1999;80:238-251.

[101] Retana, J. & Cerdá, X. 2000. Patterns of diversity and composition of Mediterranean ground ant communities tracking spatial and temporal variability in the thermal environment. Oecologia 123 pp. 436-444.

[102] Vogt, J.T., Smith, W.A., Grantham, R.A. & Wright, R.E. 2003. Effects of temperature and season on foraging activity of red imported fire ants (Hymenoptera: Formicidae) in Oklahoma. Environmental Entomology 323 pp. 447-451.

[103] Kumar, A. & O'Donnell, S. 2009. Elevation and forest clearing effects on foraging differ between surface – and subterranean – foraging army ants (Formicidae: Ecitoninae). Journal of Animal Ecology 78 pp. 91-97.

[104] Cerdá, X., Retana, J. & Manzaneda, A. 1998. The role of competition by dominants and temperature in the foraging of subordinate species in Mediterranean ant communities. Oecologia 117 pp. 404-412.

[105] Pećarević M, Danoff-Burg J, Dunn RR. Biodiversity on Broadway: enigmatic diversity of the societies of ants (Formicidae) on the streets of New York City. PLoS One. 2010;51-8. e13222. DOI:10.1371/journal.pone.0013222.

[106] King J.R. & Porter, S.D. 2004. Recommendations on the use of alcohols for preservation of ant specimens (Hymenoptera, Formicidae). Insectes Sociaux 51 pp. 197-202. doi:10.1007/s00040-003-0709-x.

[107] Andersen A, Fisher A, Hoffmann BD, Read JL, Richards R. Use

of terrestrial invertebrates for biodiversity monitoring in Australian rangelands, with particular reference to ants. Australian Ecology. 2004; 29:87-92.

[108] King J.R. & Porter, S.D. 2005. Evaluation of Sampling Methods and Species Richness Estimators for Ants in Upland Ecosystems in Florida. Environmental Entomology 34 pp. 1566-1578.

[109] Colwell, R.K., Mao, C.X. & Chang, J. 2004. Interpolating, extrapolating, and comparing incidence-based species accumulation curves. Ecology 85: pp. 2717-2727.

[110] Andrade, T., Marques, G.D.V. & Del-Claro, K. 2007. Diversity of ground dwelling ants in cerrado: an analysis of temporal variations and distinctive physiognomies of vegetation (Hymenoptera: Formicidae). Sociobiology 50 pp.1-14.

[111] Colwell, R. 2013. EstimateS: Statistical estimation of species richness and shared species from samples. Version 9. http://purl.oclc.org/estimates.

[112] Urrutia-Escobar MX, Armbrecht I. Effect of two agroecological management strategies on ant (hymenoptera: Formicidae) diversity on coffee plantations in southwestern Colombia. Environmental Entomology. 2013;42:194-203.

[113] Cerdá, X., Palacios, R. & 2009. Ant Community Structure in Citrus Orchards in the Mediterranean Basin: Impoverishment as a Consequence of Habitat Homogeneity. Environmental Entomology 38 pp. 317-324.

[114] Kindt R & Coe R. 2005. Tree diversity analysis. A manual and software for common statistical methods for ecological and biodiversity studies. Nairobi: World Agroforestry Centre (ICRAF).

[115] Porter, S.D. & Savignano, D.A. 1990. Invasion of polygyne fire ants decimates native ants and disrupts arthropod community. Ecology 71 pp. 2095-2106.

[116] Porter SD, Williams DF, Patterson RS, Fowler HG. Intercontinental Differences in the Abundance of *Solenopsis* Fire Ants (Hymenoptera: Formicidae): Escape from Natural Enemies? Environmental Entomology. 1997;26:373-384.

[117] Morrison, L.W. 2002. Long-term impacts of an arthropod-community invasion by the imported fire ant, Solenopsis Invicta. Ecology 83 pp. 2337-2345.

[118] Rivas-Arancibia, S.P., Carrillo-Ruiz, H., Arce, B.A., Figueroa-Castro, D.M. & Andrés-Hernández, A.R. 2014. Effect of disturbance on the ant community in a semiarid region of Central Mexico. Applied Ecology and Environmental Research 12 pp. 703-716.

[119] Sarty, M., Abbott, K.L. & · Lester, P.J. 2006. Habitat complexity facilitates coexistence in a tropical ant community. Oecologia 149 pp. 465-473.

[120] Parr, C.L. 2008. Dominant ants can control assemblage species richness in a South African savanna. Journal of Animal Ecology 77 pp. 1191-1198.

[121] Calcaterra, L., Cabrer, S. & Brian, J. 2016. Local co-occurrence of several highly invasive ants in their native range: are they all ecologically dominant species? Insectes sociaux 63 pp. 407-419.

[122] Dufrêne M, Legendre P. Species assemblages and indicator species: the need for a flexible asymmetrical approach. Ecological Monographs. 997;67:345-366.

[123] Podania J, Csányi B. Detecting indicator species: Some extensions of the IndVal measure. Ecological Indicators. 2010;10:1119-1124

[124] Sanabria, C., Lavelle, P. & Fonte, S.J. 2014. Ants as indicators of soil-based ecosystem services in agroecosystems of the Colombian Llanos. Applied Soil Ecology 84 pp. 24-30.

[125] Sithole, H., Tantsi, N., Bezuidenhout, H., Munyai T.C. & Munyai, L., 2020, Reconnaissance of epigeal ants at the degraded and control sites of Mountain Zebra and Mokala National Parks. Koedoe 62, a1542. https://doi.org/10.4102/koedoe. v62i1.1542.

[126] Buczkowski, G. & Richmond, D. S. (2012) 'The Effect of Urbanization on Ant Abundance and Diversity: A Temporal Examination of Factors Affecting Biodiversity. PLOS One 7.

[127] Fayle, T. M., Turner, E. C., Snaddon, J. L., Chey, V. K., Chung, A. Y. C., Eggleton, P., & Foster, W. A. 2010b. Oil palm expansion into rain forest greatly reduces ant biodiversity in canopy, epiphytes and leaf- litter. Basic and Applied Ecology, 11 pp. 337-345.

[128] Nazarreta, R., Hartke, T.R., Hidayat, P., Scheu, S., Buchori, D. & Drescher, J. 2020. Rainforest conversion to smallholder plantations of rubber or oil palm leads to species loss and community shifts in canopy ants (Hymenoptera: Formicidae). Myrmecological News 30 pp. 175-186.

Chapter 6

Biomonitoring Ecosystem: Modelling Relationship with Arthropods

Jinu Medhi, Jintu Dutta and Mohan Chandra Kalita

Abstract

Arthropods community structure and composition provides multiscale information about an environment health. Their reproduction and growth model are effective to assess the impact on ecosystem in response to stress such as anthropogenic activities (climate change) or natural (drought). Terrestrial and aquatic insects are potential bio-indicators. Terrestrial insects are an excellent model to assess the quality of terrestrial ecosystem. These insect species are assayed to detect metallic pollution and forest abundance. Soil and litter arthropods are used for examining soil quality. Honey bee mortality rates and the residues such as heavy metals, fungicides and herbicides presence in honey are good indicator of environmental pollution. The specificity of food and habitat selection by wasp population make it suitable for assessing habitat quality. Similarly butterflies habitat itself signifies a healthy ecosystem because of their sensitivity to even slightest change. Different arthropods act as keystone species and these keystone interactions also reveal many facets of an ecosystem quality. Similarly fly population such as *Drosophila subobscura* and their shift in the genetic composition indicate the global climate warming. The arthropods are explored as screening platform to understand the ecosystem resilience to disturbances. These underscores arthropods potential for evaluation of environmental impact and global climate change.

Keywords: arthropods, ecosystem, climate change, environmental pollution, biodiversity conservation

1. Introduction

The main objective of this chapter is to monitor and explore the ecosystem and biodiversity by understanding the surrounding environment and public health surveillance. By monitoring the ecosystem and biodiversity, one can assess the proven significance of ecological indicators over terrestrial and geographical scales to discover emerging changes in the structure, function, and composition of ecological system with respect to natural and man-made influences. The extent to which the ecological indicators are used in monitoring the whole ecosystem and the selection procedure followed identifying the symbiotic indicators. There have been numerous articles, books, journals, videos, and manuscripts that relate to the field

IntechOpen

of monitoring ecosystem and biodiversity. Although former researchers, educationists, and many others have contributed to this area of study with different perspectives, there are still numerous common sections untouched which seem to link most of the articles collectively.

Moreover, it has been observed that several monitoring programs have been conducted to complete the ecological indicator selection process of which arthropod ecological indicators stand stringent. The whole arthropod community structure contributes to the well-being of environment health. Arthropods reproductive system, growth, and development aids the ecologists in assessing the impact of man-made activities which stand responsible for climate change, drought and other environment- effecting disasters. Different types of arthropods are explored which helps to evaluate environmental impact and design strategies for a successful healthy ecosystem and biodiversity. Thus, the current chapter highlights on key aspects of healthy ecosystem which include arthropods, relationship between arthropods and ecosystem, concept of biomonitoring, anthropogenic factors effecting ecosystem, and analyzing biodiversity.

Arthropods are invertebrate animals with an exoskeleton, a segmented body and jointed legs. High functional, biological diversity and sensitiveness to environment, of arthropods make them suitable to be considered for utility as ecological indicators of sustainable ecosystem. The potential bioindicators groups of arthropoda include Acari, Collembola, Coleoptera, Hymenoptera and Araneae.

2. Biodiversity

In this whole universe, diversity exists at all levels of biological organization irrespective of the species ranging from macromolecules to ecosystems. The term biodiversity refers to the degree of biological variations within the ecosystem. In other words, biodiversity is the combination or blend of all biological organizations. As a whole, biodiversity refers to various forms of life existing on earth. The different forms of life on earth can be categorized into animals, plants, genes, microorganisms and the ecosystem itself. Biological resources such as species, genes, ecosystems and organisms and ecological processes of the above said resources can be manifested as biodiversity [1]. Therefore, Biodiversity is analyzed and understood at 3 major levels namely: Genetic diversity, Species Diversity, and Ecosystem Diversity.

Genetic diversity provides the genetic information of all plants, animals, species, and microorganisms with respect to the population of species. It simply deals with variety of genes within the specie and population. Species diversity is nothing but diversity at specific levels. It includes variety of species such as species richness, which refers to total number of species identified in the target area; species abundance, which refers to the relative number among the species; and taxonomic diversity, which takes into account the genetic relationship between various species. Ecosystem diversity refers to variety of habitats, ecosystems, ecological communities, and ecological processes in the environment.

2.1 Importance of biodiversity

Every single form of life on earth seems to be rare and has its own value regardless of the species under which that living organism is considered. Just like human beings all other forms of life has its own place and value. This is the right of every organism in the ecosystem. Every organism which is the part of this ecosystem has an innate right to exist in the universe regardless of its value, respect and honor.

DOI: http://dx.doi.org/10.5772/intechopen.94313

However, human beings seem to be the integral part of the natural world and hence the world value the human life more than all other living species. The environment preserves human heritage as humans are considered essential to the world. For this reason, the well-being of the coming human generations is wholly invested in the hands of existing generation. Sustaining the diversity becomes the key social responsibility of the anthropogony. Therefore, only by conserving the resources and organisms the future generation's existence can be determined. Hence, these values become ethical or moral values for preserving biodiversity.

The nature in which human beings live, grow and develop turns out to be a great enjoyment to the whole humanity. Natural environment in which human beings live plays an essential role in shaping and structuring the culture, stimulate the senses of human beings and eventually enrich our social culture. Hence, biodiversity is vested in human culture which is popularly preserved, valued, and protected. For instance, colossal amounts of money is paid by organizations in order to preserve, conserve and yield wild life as it becomes the vital part of human nature. The environment is protected, appreciated, and enjoyed only when the wild species are kept preserved. In fact, wild species enhance mankind's way of living by providing enjoyment through different types of activities such as bird watching, trekking, spotting activities, watching wildlife and so forth. The above said activities attribute aesthetic value to biodiversity.

Besides, biodiversity comes with utilitarian values that contribute to the very existence and material well-being of living organisms. Apart from emotions and feelings of human race, there are several other materialist things which provide ultimate satisfaction to a human life. This includes conservation of materials from biodiversity such as agricultural materials, source of food, clothing, medicinal values, industry materials, educational values, scientific understanding and materialistic yearnings. Thus, a rich biodiversity is essential for healthy ecosystem and stands imperative for the survival of human race.

2.2 Ecology and biodiversity

Ecology and Biodiversity are two interrelated terms that seeks to maintain, preserve, and sustain the integrity of the ecological system thorough different ways such as maintaining carbon dioxide (CO2) and oxygen balance (O2), regulating biochemical cycles, decomposing waste materials, regulating natural world climate, identifying indicators that change the environment, and provide protection services to the ecosystem. Maintaining carbon dioxide and oxygen levels in the atmosphere can be made possible only through biodiversity. Carbon dioxide accumulation in the atmosphere results in greenhouse effect which eventually leads to ozone layer depletion. Ozone layer depletion makes the earth warmer and liable to natural disasters.

Preserving biochemical cycles is another important way to maintain biodiversity levels in this ecosystem. Regulating biochemical cycles is equally important in maintaining ecological values of biodiversity. Decomposing waste materials thorough absorption and breakdown of pollutants will lead to food webs and food chains to other forms of biodiversity. Production of waste would be zero as the waste id decomposed and transformed as food to other forms of biodiversity. Thus, pollutants are broken down and absorbed naturally. The other ecological values of biodiversity include controlling and determining the natural world climate irrespective of their regions by means of influencing factors such as precipitation, temperature, and air turbulence; act as indicators of environmental change, for instance, global warming changes in ecosystem and affects crops; and protect humans from harmful weather conditions.

2.3 Threats to biodiversity

Generally, threat is defined as a natural or human-made process or event responsible for causing adverse effects on the sustainable use of biodiversity components at large. The biological diversity and wealth of our ecosystem has been rapidly decreasing due to the clearly pointed anthropogenic activities. Several studies have discussed the disappearance of large number of species along with the possible threats to the species as well as ecosystems. In the recent times, highest number of threats to ecosystems has been recorded. The reason for the failure of ecosystem is identified as human mismanagement of biological resources often stimulated by misguided policies which results in loss of biodiversity. Loss of biodiversity can further lead to decline in ecosystem process, decline in plant production, and lowered resistance to environmental pollutant such as physical, chemical and biological. The quicker rates of species extinctions that the world is facing now are largely due to human activities. Given below are the major threats to biodiversity:

2.3.1 Habitat destruction

Increased voracious demand for resources results to use of land species which eventually acts as cause to loss of genetic diversity, changes in ecosystem such as disease outcrops, population increase or decrease, habitat fragmentation, and reduction in the number of species of ecosystem. The above mentioned reasons lead to heavy biodiversity loses. Habitat destruction, therefore, becomes a threat to biodiversity.

2.3.2 Overexploitation of biological resources

When individuals of particular species which can be sustained for a longer period of time with its reproductive capacity are decomposed at higher rate, population is said to be harvest or exploited at higher rate. For instance, human-induced activities such as fishing, hunting, food gathering, trade and so forth are responsible for overexploitation of higher sustainable biological resources. Over exploitation leads to extinction of biological resources and thereby reduces the number of species in the ecosystem. Exploitation of resources with the consent of law is termed as cropping. While exploitation of species even after providing protection is termed as poaching. Thus, overexploitation of resources or biological resources will turn out to be a major threat to biodiversity.

2.3.3 Pollution

Any kind of pollution, be it physical, chemical, biological or thermal is a hazard to biodiversity. Majority of the species living in their habitats are prone to harmful industrial activities, pollution, and excessive use of chemicals. This kind of pollution eventually harms the ecosystem.

2.3.4 Biological invasions

Biological Invasion can be intentional or accidental. Changes within the ecosystem are mainly due to the biological invasion of new species. The newly introduced classes are organisms that arise from habitats in which they were not found. These new introduced species from new habitats are generally termed to as biological pollutants. The ecological impacts of biological invasion include disorder of food webs, out competition, hybridization, disorder of ecosystem, disease transmission, plant pathogenic influences, and extinction of species in peculiar situations. The new species may be introduced intentionally for different types of reasons such as ornamental concerns; agriculture; hunting and spotting activities; biotechnology for scientific research; and trade.

2.3.5 Climatic changes

Climate change is one of the greatest concern especially when global carbon dioxide increases in the atmosphere resulting to global warming. Economic stability of an ecosystem is tolerated when most species originate within a narrow physiological limit. Hence, changes may be gradual or abrupt and hence result in species extinction.

2.3.6 Population

With increasing human population, vigorous demand for raw materials also increases which results in changes in biodiversity. Hence, controlling human population will be the only solution to conserve biodiversity for the coming generations.

2.4 Biodiversity conversation

Biodiversity Conversation embodies maintenance, preservation, conservation, and enhancement of crucial biological diversity components. Conservation is referred to as the sustainable use of biological resources with the aim of protecting them for the coming generation and at the same time protect them exploitation. Preservation is keeping the materials in a safe manner without altering it. Biodiversity Conservation and Sustainable Development are interlinked to meet the needs of current generation without undermining the thought of preserving for the coming generations to meet their basic needs. In other words, it establishes a balance between the ecosystem and living organisms which ensures biodiversity.

The narrow practical arguments for biodiversity conservation are several. For instance, human being spring multiple benefits from nature such as pulses, cereals, firewood, fruits, construction material, fiber, industrial products such as dyes, lubricants, tannins, perfumes, and resins; and medical products. More than twenty five of the drugs sold in the market are derived from plants and twenty five thousand species of plants are used to prepare traditional medicines that are used by human beings around the world. Hence, all the species in the ecosystem depend on each other.

The broad practical argument claims that biodiversity plays a major role in ecosystem services. Through photosynthesis, Amazon forest produces twenty percent of the whole oxygen in the earth's atmosphere. While the ethical argument for conserving biodiversity to plant, animal and microorganisms living in this planet. Philosophically or spiritually, every species has an intrinsic value. Hence their well-being should be taken care and pass on biological inheritance to the coming generations.

2.5 Concept of biomonitoring

The concept of monitoring can be defined as the process of observing and measuring the state of key indicators such as physical, chemical and biological with respect to the element of environment or the medium. Physical monitoring is usually carried out considering the physical parameters such as temperature, climate, and other variables. Chemical monitoring monitors the chemical variables responsible for environmental pollution. However, these two types of monitoring failed to come up with the long-term effects of pollution on the environment. Assessing the ecosystem by taking into account the physical and chemical monitoring seems to be unreliable. Hence, the concept of biomonitoring has been introduced to assess the long-term effects of identified pollutants on ecosystem and produce reliable results. Biomonitoring or Biological monitoring is introducing biological variables to assess the structural, functional and compositional aspects of an ecosystem. These biological variables play an important role in controlling environmental alterations.

Biomonitoring is a systematic use of symbiotic organisms to assess the quality of environment. It enables to check for the additive effects of pollutants and monitors the overall health condition of ecosystem [2]. Hence, biomonitoring acts as a supplement to physical and chemical monitoring techniques that are commonly applied. Biomonitoring is defined as an act of observing, noticing, and assessing the changes within the ecosystem, structure of ecosystem, composition of biodiversity, and functions of ecosystem and biodiversity including different types of natural habitats, keystone species and population [2]. The advantages of biomonitoring are rich than those of physical and chemical monitoring which include: (1) it reflects the overall environmental integrity comprising of physical, chemical and biological monitoring; (2) it imparts an integrated and holistic measure of ecological condition by uniting stresses over a period of time; and (3) it provides a better understanding of healthy environment to the public surveillance than others.

Biomonitoring can be achieved by bringing about a change in the structural, functional and compositional aspects of biodiversity and ecosystem that are affected due to the adverse anthropogenic activities. Different parts of the world are conducting programs with the aim of spreading awareness on pollutants to the public. Several approaches, methods and strategies are identified to monitor the ecological pollutant out of which four of them have been marked approval. First approach is to monitor the effects of pollution depending on the absence or presence of taxa, changes in its composition or any other drastic changes. In other words, the first step is to monitor the adverse effects of pollutant within the community. After this, concentration of pollutants in indicators should be measured. Later, effects of identified pollutants on organisms should be assessed and classify them to abiotic and biotic indicators. Lastly, identify different strains of species which develop resistance in response to a pollutant.

Several micro-organisms are used as bioindicators to assess the impact of pollutants on the whole ecological system. Some of the bioindicators include protozoa, fishes, algae, macroinvertebrates and microinvertebrates [3]. According to Nesemann [4], arthropods is one of the macroinvertebrates that seems to be dominating seabed groups found in the seafloor [4]. Of all the other macroinvertebrates, arthropods are found to be more dominant bioindicator with respect to ecological pollution tolerant, followed by molluscs and annelids.

According to Holt and Scott [5], bioindicators are the species, communities and processes that are used to assess the environmental quality and record the changes happening in the ecosystem over a period of time [5]. Generally, environmental changes are often interconnected and interlinked with man-made disturbances such as pollution, droughts, climatic changes, and so forth. These bioindicators expanded their arena to all types of environments such as aquatic and terrestrial. Past studies claimed that bioindicators successfully indicates the condition of the environment along with the rate of tolerance to environmental variability. It is also observed that species or indicators with low or narrow tolerance act sensitive to the changes occurring in the ecosystem. In contrast, indicators or species with high or broad tolerance act less sensitive to the changes occurring in the environment which disturb the community.

To conclude, biomonitoring and bioindicators are more or less the same and hence can be interchangeable within the science community with slight difference. While bioindicators assesses the impact of environmental pollutants qualitatively, biomonitors determine the responses of pollutants quantitatively. Therefore, the main functions of bioindicators or biomonitoring include: (a) monitor or assess the environment; (b) monitor or assess ecological process; and (c) monitor or assess biodiversity. Thus, biomonitoring lies at the core of ecosystem and has become the essential and effective tool to study environmental exposure to chemicals, pollutants, and other hazardous materials. Biomonitoring studies measures the responses and recoveries of water communities from the ecological disturbances, environmental pollution, and evaluate the relationship between physical, chemical and biological components.

3. Arthropods as ecological indicators

The important components of the ecosystem that occupy vital positions in food webs, changing population, and communities are called as Arthropods. Arthropods play multiple roles in this ecosystem such as going about as herbivores, decomposers, predators, parasites, seed dispersers, and pollinators [6]. The peculiar characteristics of arthropods such as small body size, high diversity, increased reproductive capacity, easy sampling, and less sensitive to environmental changes make them suitable for environmental biomonitoring. For these reasons, arthropods are used as biological indicators to monitor and assess the impact of pollutants on the ecosystem.

Usually, arthropods are used as bioindicators for the following reasons: (a) the most frequently polyphagous predators that play a crucial role in biological control; (b) groups are made easily with danger traps; (c) allow elevated statistical analysis. According Da Rocha [7], environmental indicators such as physical, chemical, human, and biological shows changes in the ecosystem. These indicators should be analyzed in the complex dynamics of the environment.

Biological indicators give insight of biological systems. These indicators provide significant information for prioritizing conservation areas and come up with better ecosystem management plans. Arthropods possess explicit spatial and temporal scales that distinguishes high patch sizes, patch dynamics, quick turnover, geographic distributions, larger population size when compared with birds and vertebrates. Thus, arthropods could be used reliably to infer ecosystem function and habitat condition.

Arthropods occupy the widest diversity of microhabitats and niches, and play more ecological roles, than any other group of animals. They have diverse body sizes, agilities, and growth rates. Arthropods have been recognized as efficient indicators of ecosystem function and recommended for use in conservation planning and many researchers have assessed habitat quality and measured habitat differences using arthropods [7]. Important biondicator groups include Coleoptera (Carabidae, Curculionidae, Staphylinidae), Collembolla (Springtails), Diplura, and Hymenoptera (Formicidae). *Parallelomorphus laevigatus* is good indicator of toxic elements. Arthropod groups have been used to track global health of ecosystem in many contexts. Major arthropods and their taxonomic classification are represented in the **Figure 1**. Ecological indicators such as arthropods help to identify the impacts of natural and anthropogenic disturbances on biota.

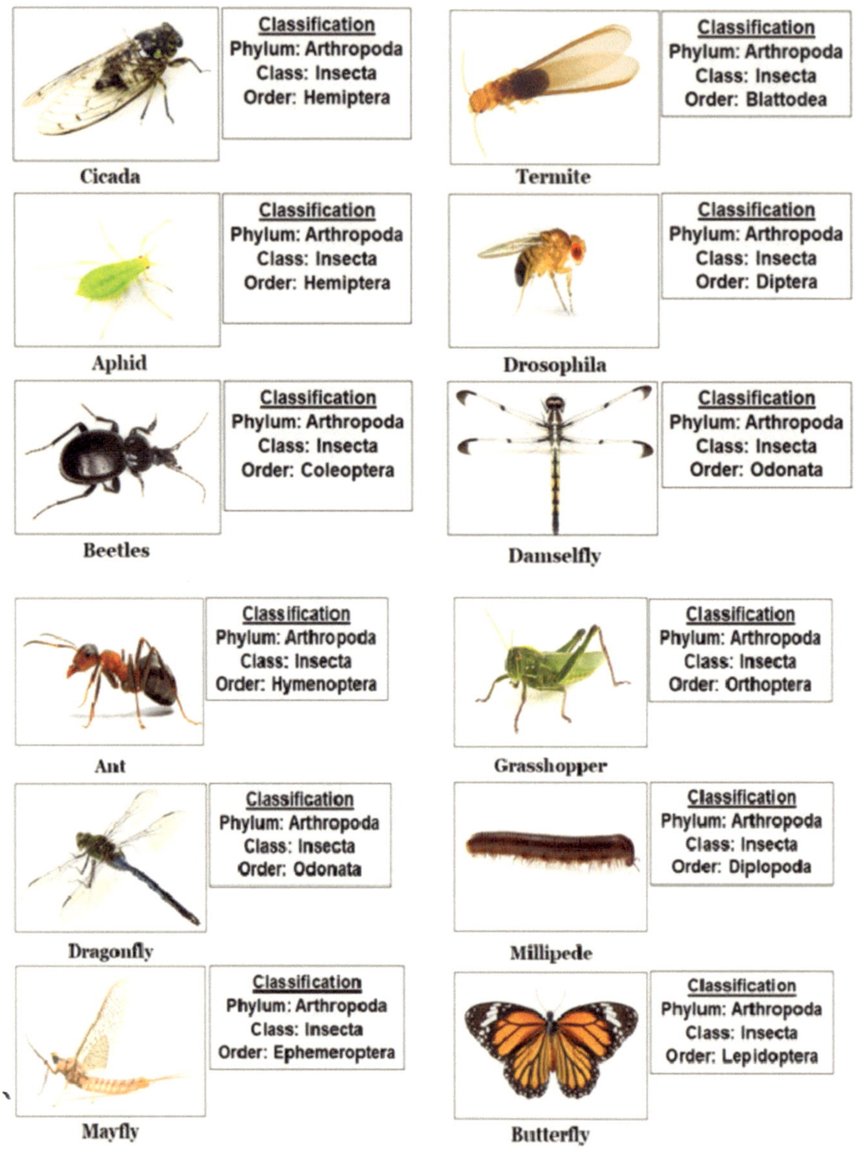

Figure 1.
Major arthropods and their taxonomic classification.

Among the arthropods, it is indicated that insects becomes useful as they represent half of the species and allows to assess different habitats in a refined manner.

3.1 Arthropod characteristics as good indicator

The attributes that make Arthropod biomonitoring acceptable because of the following attributes (1) cost-effective (2) their easy, reliable identification (3) respond differently to disturbance regimes [8]. Arthropods are commonly sampled from different habitats Spiller et al. [9]. Soil Arthropods are biological indicators of farm and plantation ecosystem processes because their existence regulate nutrient dynamics, soil quality, and are useful to reveal ecosystem condition [10, 11]. Some arthropod bioindicator groups live, feed and reproduce in the soil. They are highly sensitive to soil quality alterations [12, 13]. Moreover the high functional and biological diversity of arthropods provide evidence for their utility as ecological indicators of sustainable forest [14].

3.2 Arthropod characteristics as bad indicator

Studies reported the disadvantages of Diptera species (flies) such as great taxonomic difficulty, especially in their larval stage is an important barrier to its use as environmental bioindicators [15]. Challenges such as sampling methods and the proper identification of soil and litter arthropod diversity up to species level demands further research to overcome the disadvantages for utility of soil arthropods while assessing soil quality. An investigation to the most suitable methods for sampling soil and litter arthropods will be beneficial to strive arthropod potential for biomonitoring [16].

3.3 Cicadas

Cicadas are large hemipteran insects characterized by unique life-history traits. Their interesting attributes are follows (1) extraordinarily long life cycles (2) a subterranean/terrestrial habitat transition (3) xylem sap-feeding (4) melodious sound production.

Cicada fauna is known for community responses to climate change. Different studies taken up this model to understand its ecophysiological consequences to climate change. The cicadas are promising candidates for use as bioindicator species to monitor ecological impacts of climate change. Recently cicada nymphs have been studied in response to emerging novel environmental stress such as construction and demolition (C&D) waste. Studies carried out in uncontaminated and contaminated habitats by C&D waste where the cumulative effects on fitness and community structure of cicada nymphs such as biodiversity, community structure, population dynamics and morphology were investigated. A significant negative response was reported in *Cryptotympana atrata* and *Platypleura kaempferi* including higher ratio of malformed nymphs [17]. Findings also identified soil compaction due to urbanization as a key reason to cicada diversity loss. The response of cicada nymphs to C&D waste have significant implications for the habitat destruction. Cicada nymphs may be suitable bioindicators for underground-habitat-quality monitoring. However further research is required to reveal the association between the magnitude of C&D waste contamination with the fitness and population dynamics of cicada nymphs.

3.4 Aphids

Aphids are known for their feeding style and close association with host plants. Due to various environmental stresses, developmental instability in morphological characters such as fluctuating asymmetry can occur in aphids. Heavy metal accumulation, pesticide application, other pollutants and anthropogenic disturbances posses (**Figure 1**).

3.5 Beetles

Beetles form a large group of organisms which differ taxonomically and ecologically. The attribute that makes beetles distinguishable from insects are presence of hardened fore wings which provides protection to membranous hind wings. They feed on animal waste, rotten wood, animal carcasses and make soil suitable for vegetation. Beetles are very sensitive to environmental modifications and can be cost effectively sampled by employing different methods and these criteria makes beetles an excellent model for monitoring terrestrial ecosystem [18].

Beetles are used to detect environmental contaminants including biomonitoring of metal in the field studies. Biomonitoring programs include measurements of metal in these invertebrates. *Carabus lefebvrei*, is considered suitable for evaluation of As and Hg in the environment because of high bioaccumulation factor. Similarly a positive correlation is reported between body mass and Pb in *Notiophilus biguttatus*, and *Notiophilus rufipes* and *Calathus melanocephalus*. Sex specific variation in the content of Zn has been reported in carabids including *Poecilus cupreus*, *Pterostichus melanarius*, *Pterostichus niger*, *Pseudophonus rufipes*, *Carabus nemoralis* and *C. granulatus*.

Carabid beetles diversity is suitable for studying ecological impacts of anthropogenic activities. They are extremely sensitive to increasing human disturbances. Their abundance in grasslands and boreal forest was studied in relevance to habitat disturbance gradient. Staphylinid beetles biotopes have been considered for various land management practices.

According to Spector [19], dung beetles are found in different types of landforms such as forests, grasslands, deserts and grasslands [19]. In addition to, dung beetles feed on fungi, fruits, decomposing leaves. Taking these characteristics into account, dung beetles are considered as an ideal indicator to monitor biodiversity. Out of all bioindicators, dung beetles are utilized for clear-cutting, fragmentation, forest modification, fragmentation, and logging in the tropical regions [20]. Several aquatic insects groups or arthropods are identified as aquatic environment bioindicators. Hydrophilid beetles such as *Cercyon unipunctatus* are important indicators of aquatic pollution.

Beetles from order Coleoptera and Family Carabidae are important predators. They participate in biological control, biological monitoring of pollution from oil, sulfur, herbicides, CO2, insecticides and radioactive phosphorus [7]. Predatory aquatic beetles are good indicator of trace elements. These are good candidates to monitor the trend of metal accumulation in aquatic invertebrates thereby making it suitable for distinguishing an impacted or non-impacted environments [21].

Dragonflies or Odonata species act sensitive to changes occurring within their habitat especially flooded areas, lakes and drainage areas. Several other species of the families such as Coleoptera, Heteroptera, Gyrinidae, Dytiscidae, Notonectidae, Plecoptera and Ephemeroptera have high adaptive capacity, dominating capacity, reflect ecological and geographical changes, and their biodiversity conservation. Thus, the tolerance of aquatic organisms to metals are found to be less, however, tolerates toxic agents responsible for environmental stress.

3.6 Termites

The termites are detritivores and feed on dead and decaying organic matter. Hence they are efficient nutrient recyclers which colonize dead and decaying organic matter. Termite mounds are considered as 'hotspots of fertility' or 'nutrient patches'. These increases plant and animal diversity in the ecosystem. The population dynamics and species richness in termites can be used as an environmental bio indicator to detect habitat disturbance.

Termites possess several characteristics that make them appropriate to use as bioindicators of habitat quality. The richness, abundance and composition of termite communities analyzed in vegetations with different levels of anthropogenic disturbances [22]. The ecological behavior of termites is affected by land use change and disturbance level. According to the land use types and disturbance level, a variation in termite species' richness and evenness, relative abundance, and biomass of termite were reported. Hence a major factor for declining termite diversity is found to be habitat disturbance [23]. The conformity between environmental variables and ecological data can effectively model termite communities as potential tools for ecological monitoring.

India's Coffee forests are considered as self-sustaining ecosystem and inter linked with various biotic partners including termites' community. Termite mounds act as important bioindicator that reveals ecology of the region. Their distribution and abundance may provide vital clues with respect to nutrient recycling and soil dynamics inside the coffee ecosystem.

3.7 Flies

The ecology of *Drosophila* species Drosophilid (Diptera) and their assemblage has the potential as bioindicator in open environments. Family Syrphidae is widely known for its well-known taxonomy and the larvae require different environmental condition and these characteristics makes these flies a potential bioindicator. Similarly Sarcophagidae flies are considered as good bioindicator of environmental pollution by heavy metal, fibre asbestos and waste chemicals.

3.8 Damselflies

Damselflies (Zygoptera) are insects belonging to the order Odonata and their nypmhs spend most of their lives as aquatic nymphs. Damselfly larvae are sensitive to water depth, water movement, and pH. Fluctuating asymmetry in damselflies has been used widely to investigate the effects of environmental pollution. They are considered moderately sensitive to pollution. Damselflies together with other macroinvertebrates considered as common bioindicators of stream and wetland health.

3.9 Ants

Ants are eusocial insects from Formicidae family and their communities headed by queen or queens. Worker ants are wingless females which carry out activities such as foraging, take care of queens offspring, they live in structured nest underground.

Studies suggest, ants are extensively used as effective bioindicators that hold responsible for ecosystem management and biodiversity restoration [24]. Ants seem to have high sensitivity to environmental disturbances such as grazing, forest fires, forest conversion, forest thinning, forest fragmentation, species invasion, and other forms of disturbance [25].

They are considered as important part of soil macro fauna because of their ability to restore soil quality. These bioindicators are required to monitor adverse changes in soil quality and can provide warning. Ants-soil quality model can be explored to identify sustainability of soil resource [26].

The biogeography of ants community structure has been used for validation. Ants are used to check habitat disturbance. Their composition have been used to identify ecological change in different habitats around the world including rainforests of Australia, Brazilian Savanna, Shivalik Mountains of Himalayas [27, 28]. They have been used as representatives of ecological trend in the mining site [29].

3.10 Dragonflies (Anisoptera)

Land used intensification can be studied by evaluating Odonata species richness, body size and individual species' response. An Odonates body size variable found to better variable than richness to tract integrity of original vegetation [30]. Methylmercury (MeHg) levels in dragonfly larvae and water were measured. In aquatic systems dissolved MeHg concentrations levels in dragonfly larvae are useful indicators [31]. Their sensitivity to habitat quality and the amphibious life cycle make Dragonflies (Anisoptera) an efficient environmental tool to track micro changes within the confines of coffee ecosystem. Similarly these Dragonflies are best suited for evaluating water quality and any environmental changes in a coffee ecosystem.

3.11 Mayflies

Mayflies are found in a wide variety of habitats and are very sensitive to pollution. They are considered as valuable indicator of water pollution [32]. Ephemeroptera larvae are recognized as bioindicators and used in many monitoring programmes for their sensitivity to oxygen depletion in running waters [33, 34]. They are considered as keystone species. Different mayfly genera such as *Tricorythopsis* (Leptohyphidae) and *Camelobaetidius* (Baetidae), are considered as potential bioindicators of different anthropogenic activities [35]. Along with caddisflies and stoneflies, mayflies are one of the three most commonly used indices to understand aquatic ecosystem health [32]. The quantitative biological information for conserving and managing freshwater ecosystems.

3.12 Grasshoppers

Grasshoppers are a dominant group of herbivorous insects and their diversity, sensitivity to disturbances and ease of sampling make them quality bioindicators for land management. Grasshopper assemblage dynamics is considered reflective to human land use.

They are sensitive to disturbances and their ease of sampling makes it a good model for land management studies. The accumulation of metal such as accumulation patterns for Cd, Ni, Cr, Zn, and Cu as result of industrial effluents, agricultural runoff, vehicular smoke, domestic and sewage wastes, and use of fertilizers studied in acridid grasshopper (*Oxya hyla hyla*) [36]. Studies also suggest *Serpusia opacula as useful indicator for anthropogenic activities* [37].

3.13 Millipedes

Millipedes are involved in breakdown of organic matter and because of their sensitivity to habitat change considered as important bioindicator taxa. Biodiversity conservation efforts should consider these invertebrates. The content of some elements such as Ca, P, K, Mg, Na, Fe, Cu and B in adult bodies of *Glomeris hexasticha* millipedes reflect the intensity of environmental pollution.

3.14 Butterflies

Butterflies with a lifespan ranging from 15 to 30 days, are considered to be most potential bioindicator group because (1) easily identifiable by DNA barcoding (2) require little labour to collect (3) maintain symbiosis with definite host plant. They respond extremely quickly to environmental changes and acts as early warning system of biodiversity reduction. A decline in butterflies can suggest the richness of other British species, in particular birds such as blue tits, jays and sparrows.

According to Halder [38], **butterflies** are used as bioindicators of robust ecosystems as they have firm connections with habitat variants such as meadows, hilly regions, and edges of woodlands, flower-filled fields, sunny conditions, and plenty of herbaceous plants [38]. Monitoring butterfly abundance indicates the presence of seminatural conditions. These seminatural conditions such as, flower abundance, understory herb cover, and vegetation diversity promotes butterfly diversity in an ecosystem [38–40]. Kitahara [40] claimed that richness of butterfly species is associated with nectar plant species richness, vascular plant species richness, and herbaceous plant species richness [40]. The results of the study claimed that even in conifer plantations, butterfly conversation should be maintained in forestry practices [38, 40].

Butterfly often live in close association with specific larval host plants and carry out different pollinating activities. This attributes makes butterflies a strong indicators of the presence of particular plant taxa. The composition of different butterfly communities in a specific habitat suggest environment quality and its ability to support a diverse arthropod community [41]. The rare butterfly *Tomares nesimachus* (Lycaenidae) serve as umbrella species and their abundance is dependent on different ecological factors [42]. Its population dynamics is suggested to be good indicator of ecosystem functioning.

In addition to, moths are the bioindicators widely used during vegetation recovery as a result of environmental disturbance. Moth families such as Arctiinae, Catocalinae, Heliothinae, Noctuinae, Hermeniidae, and Phycitinae respond positively to the environmental disturbances, while others such as Ennominae, Geometrinae, Epipaschiinae, Lymantriidae, and Anthelidae respond negatively. For the above reasons, moths are considered as effective bioindicators. These different responses to environmental changes make them suitable bioindicators.

4. Conclusion

From the study it can be claimed that appropriate use of biological indicators is fundamental for biomonitoring or environmental monitoring. The primary features of bioindicators include species richness and diversity, indicators can be handled easily, showcase high faithfulness towards ecology, more sensitive and fragile to ecological changes. It is observed that some of the environment species in the ecosystem tend to respond in better ways to the changes in the environment. Odonata species are observed to be highly sensitive to environmental changes occuring in the water. While some other species such as Plecoptera, Heteroptera, Coleoptera, and Ephemeroptera are highly adaptive in nature. With respect to land insects, Coleoptera Order has several bioindicators. Different types of bees are used to monitor trace metals in pesticides, herbicide effects, ecological conditions, and radioactivity. Therefore, this study concludes that arthropods are environmental bioindicators which monitors, assesses and maintains a healthy biodiversity conservation with a healthy ecosystem.

Author details

Jinu Medhi[1*], Jintu Dutta[2] and Mohan Chandra Kalita[1]

1 Department of Biotechnology, Gauhati University, Guwahati, Assam, India

2 Centre for the Environment, Indian Institute of Technology Guwahati, Assam, India

*Address all correspondence to: jinumedhi@gmail.com

References

[1] Mutia TM. Biodiversity conservation. *Short Course IV on Exploration for Geothermal*. Resources. 2009:1-22

[2] Derocles SA et al. *Advances in ecological research* Vol. In: 58 1-62. 2018

[3] Sharma K, Chowdhary S. Macroinvertebrate assemblages as biological indicators of pollution in a Central Himalayan River, Tawi (JK). International Journal of Biodiversity and Conservation. 2011;**3**:167-174

[4] Nesemann H, Sharma G, Sinha R. Benthic macro-invertebrate fauna and" marine elements" sensu Annandale (1922) Highlight the valuable dolphin habitat of river Ganga in Bihar-India. *TAPROBANICA: The Journal of Asian Biodiversity*; 2011

[5] Holt, E. & Miller, S. (2011) Bioindicators: Using Organisms to Measure.

[6] Maleque MA, Ishii HT, Maeto K. The use of arthropods as indicators of ecosystem integrity in forest management. Journal of Forestry. 2006:113-117

[7] da Rocha JRM, De Almeida JR, Lins GA, Durval A. Insects as indicators of environmental changing and pollution: a review of appropriate species and their monitoring. Holos environment. 2010;**10**:250-262

[8] Pearce JL, Venier LA. The use of ground beetles (Coleoptera: Carabidae) and spiders (Araneae) as bioindicators of sustainable forest management: a review. Ecological Indicators. 2006;**6**:780-793

[9] Spiller MS, Spiller C, Garlet J. Arthropod bioindicators of environmental quality. Revista Agro@mbiente On-line. 2017;**12**:41-57

[10] Suheriyanto, D., Zuhro, Z., Farah, I. & Maulidiyah A. in *Journal of Physics: Conference Series.* 012180 (IOP Publishing). 2019

[11] Walker JT, Suckling DM, Wearing CH. Past, present, and future of integrated control of apple pests: the New Zealand experience. Annual Review of Entomology. 2017;**62**:231-248

[12] Parisi V, Menta C, Gardi C, Jacomini C, Mozzanica E. Microarthropod communities as a tool to assess soil quality and biodiversity: a new approach in Italy. Agriculture, Ecosystems & Environment. 2005;**105**:323-333

[13] Parisi V, Menta C. Microarthropods of the soil: convergence phenomena and evaluation of soil quality using QBS-ar and QBS-c. Fresenius Environmental Bulletin. 2008;**17**:1170-1174

[14] Langor DW, Spence JR. Arthropods as ecological indicators of sustainability in Canadian forests. The Forestry Chronicle. 2006;**82**:344-350

[15] Frouz J. Use of soil dwelling Diptera (Insecta, Diptera) as bioindicators: a review of ecological requirements and response to disturbance. Agriculture, Ecosystems & Environment. 1999;**74**:167-186

[16] Nsengimana, V. Use of soil and litter arthropods as biological indicators of soil quality in forest plantations and agricultural lands: A Review. (2018).

[17] Hou Z, Liu Y, Wei C. Influence of construction and demolition waste on fitness and community structure of cicada nymphs: New bioindicators of soil pollution. PLoS One. 2018;**13**:e0203744

[18] Ghannem S, Touaylia S, Boumaiza M. Beetles (Insecta: Coleoptera) as bioindicators of

the assessment of environmental pollution. Human and Ecological Risk Assessment: An International Journal. 2018;**24**:456-464

[19] Spector S. Scarabaeine dung beetles (Coleoptera: Scarabaeidae: Scarabaeinae): an invertebrate focal taxon for biodiversity research and conservation. The Coleopterists Bulletin. 2006;**60**:71-83

[20] Nichols E, Larsenb T, Spectora S, Davise AL, Escobarc F, Favilad M, et al. Global dung beetle response to tropical forest modification and fragmentation: a quantitative literature review and meta-analysis. Biological Conservation. 2007;**137**:1-19

[21] Burghelea CI, Zaharescu DG, Hooda PS, Palanca-Soler A. Predatory aquatic beetles, suitable trace elements bioindicators. Journal of Environmental Monitoring. 2011;**13**:1308-1315

[22] Viana-Junior AB, Reis YT, Costa APM, Souza VB. Termite assemblages in dry tropical forests of Northeastern Brazil: Are termites bioindicators of environmental disturbances? Sociobiology. 2014;**61**:324-331

[23] Pribadi T, Raffiudin R, HARAHAP IS. Termites community as environmental bioindicators in highlands: a case study in eastern slopes of Mount Slamet, Central Java. Biodiversitas Journal of Biological Diversity. 2011;**12**

[24] Underwood EC, Fisher BL. The role of ants in conservation monitoring: if, when, and how. Biological Conservation. 2006;**132**:166-182

[25] Sinclair J, New T. Pine plantations in south eastern Australia support highly impoverished ant assemblages (Hymenoptera: Formicidae). Journal of Insect Conservation. 2004;**8**:277-286

[26] De Bruyn LL. *Invertebrate Biodiversity as Bioindicators of Sustainable Landscapes*. Elsevier; 1999. pp. 425-441

[27] Lawes MJ, Moore AM, Andersen AN, Preece ND, Franklin DC. Ants as ecological indicators of rainforest restoration: Community convergence and the development of an Ant Forest Indicator Index in the Australian wet tropics. Ecology and Evolution. 2017;*7*(20):8442-8455

[28] Ribas CR, Campos RB, Schmidt FA, Solar RR. Ants as indicators in Brazil: a review with suggestions to improve the use of ants in environmental monitoring programs. Psyche. 2012;*2012*

[29] Cooper C. *Simplifying the use of ants as bioindicators on mine sites*. University of York; 2018

[30] Rocha-Ortega M, Rodríguez P, Córdoba-Aguilar A. Can dragonfly and damselfly communities be used as bioindicators of land use intensification? Ecological Indicators. 2019;**107**:105553

[31] Jeremiason JD, Reiser T, Weitz R, Berndt M, Aiken GR. Aeshnid dragonfly larvae as bioindicators of methylmercury contamination in aquatic systems impacted by elevated sulfate loading. Ecotoxicology. 2016;**25**:456-468

[32] Voshell, J. R. (2002). *A guide to common freshwater invertebrates of North America* (No. Sirsi) i9780939923878.

[33] Menetrey, N., Oertli, B., Sartori, M., Wagner, A. & Lachavanne, J. (2007). *Pond Conservation in Europe* 125-135 .

[34] Zedková B, Rádková V, Bojková J, Soldán T, Zahrádková S. Mayflies (Ephemeroptera) as indicators of environmental changes in the past five decades: a case study from the Morava and Odra River Basins (Czech

Republic). Aquatic Conservation: Marine and Freshwater Ecosystems. 2015;*25*(5):622-638

[35] Firmiano KR, Ligeiro R, Macedo DR, Juen L, Hughes RM, Callisto M. Mayfly bioindicator thresholds for several anthropogenic disturbances in neotropical savanna streams. Ecological Indicators. 2017;*74*:276-284

[36] Azam I, Afsheen S, Zia A, Javed M, Saeed R, Sarwar MK, et al. Evaluating insects as bioindicators of heavy metal contamination and accumulation near industrial area of Gujrat, Pakistan. BioMed research international. 2015

[37] AGRIPPINE Y-FJ, DIDIER MA, XU S-Q. Diversity, abundance and distribution of grasshopper species (Orthoptera: Acrididea) in three different types of vegetation with different levels of anthropogenic disturbances in the Littoral Region of Cameroon. Journal of Insect Biodiversity. 2020;**14**:16-33

[38] Halder I, Barbaro L, Corcket E, Jactel H. Importance of semi-natural habitats for the conservation of butterfly communities in landscapes dominated by pine plantations. Biodiversity and Conservation. 2008;(**5**):1149-1169

[39] Bergman K-O, Ask L, Askling J, Ignell H, Wahlman H, Milberg P. Importance of boreal grasslands in Sweden for butterfly diversity and effects of local and landscape habitat factors. Biodiversity and Conservation. 2008;**17**:139-153

[40] Kitahara M, Yumoto M, Kobayashi T. Relationship of butterfly diversity with nectar plant species richness in and around the Aokigahara primary woodland of Mount Fuji, central Japan. Biodiversity and Conservation. 2008;**17**:2713-2734

[41] Miller DG III, Lane J, Senock R. Butterflies as potential bioindicators of primary rainforest and oil palm plantation habitats on New Britain, Papua New Guinea. Pacific Conservation Biology. 2011;**17**:149-159

[42] PE'ER G, Settele J. The rare butterfly Tomares nesimachus (Lycaenidae) as a bioindicator for pollination services and ecosystem functioning in northern Israel. Israel Journal of Ecology and Evolution. 2008;**54**:111-136

Chapter 7

Vertical Arthropod Dynamics across Organic Matter Fractions in Relation to Microclimate and Plant Phenology

María F. Barberena-Arias and Elvira Cuevas

Abstract

Plant diversity is a key factor influencing belowground dynamics including microclimate and decomposer arthropod communities. This study addresses the effect of individual plant species on belowground arthropods by focusing on seasonal variations in precipitation, temperature and arthropods along the vertical organic matter profile. In the Guanica Dry Forest, Puerto Rico, microclimate was described and 5 plant species and 10 trees/species were selected. Under each tree, for one year, temperature was measured and samples collected along the organic matter fractions. Collected arthropods were standardized to ind/m^2, identified to Order/Family and assigned to morphotypes. The annual temperature pattern was similar for all species and OM fractions. Arthropod abundance was similar among plant species and higher in humus than in litter fractions. Richness and species composition were different among plant species and OM fractions. All plant species and OM fractions showed low arthropod abundance and richness, and similar arthropod species composition in the dry season, while in the wet season abundance and richness were higher and species composition varied across plant species and OM fractions. These data suggest that arthropods form specific assemblages under plant species and stages of decomposition that, during the dry season, represent a subgroup adapted to extreme environmental conditions.

Keywords: Berlese funnels, decomposition, extreme conditions, plant structure, seasonal dynamics

1. Introduction

Individual trees modify the below microclimate, resources and associated biodiversity [1]. Examples of trees modifying the below microclimate include trees serve as wind shelters [2], canopies intercept rainfall [3] and solar radiation [2, 4–6] resulting in calm drier and warmer below microclimate in comparison to nearby open areas. Through leaf fall, trees influence belowground microclimate and litter quantity and quality, for example dark litter warms more than light-colored litter [7], deciduous species show a pulse in litter production [8] and litter C: N and C:P ratios directly influence soil N and P [9] and N transformation rates [10]. In addition, physicochemical changes that happen during litter decomposition [11] are paralleled in the vertical stratification of the organic matter (OM) [12]. As a consequence, trees produce spatial variation in microenvironments and a patchy distribution of litter that is vertically stratified into progressive decomposition stages [13].

The distribution of trees results in a patchy distribution of litter and associated organisms [14–16]. For example, microarthropods were more abundant in aspen than in pine [17] earthworms were abundant under *Qualea* and completely absent under *Dicorynia guianensis* [15] and *Heliconia caribaea* [18]. On the other hand, OM in early decomposition was dominated by cellulolitic bacteria, omnipotent fungi and collembolans while later on, by proteolytic bacteria, no-lygnolitic fungi, diplopoda and isopoda [19]. Cryptostigmata and collembolans dominated late decomposition [20], while others report Cryptostigmata diversity to be independent of the decomposition stage [21] but mites and collembolans species composition changed simultaneously with a decrease in collembolan abundance as decomposition proceeded [22]. In Guanica, enzymatic activity and microbe diversity were consistently different between *Tabebuia*, an evergreen facultative deciduous species, and *Ficus* and *Pisonia*, two deciduous species [23]. In this forest, arthropod species composition was different among tree species, and this difference was better explained by detritivores that covaried with the physicochemical characteristics of mature green leaves [24].

With the overwhelming impact of human activities on biodiversity [25, 26], an enormous amount of studies (e.g. [25, 27–30] are only few examples) have addressed the relationship between diversity and ecosystem processes at the stand and local scales, but few happen at the individual tree species scale [23, 24, 31]. Therefore, we evaluated how aboveground diversity, specifically individual tree species, modify microclimate and litter, and the relation to the dynamics and diversity of belowground arthropods in the vertical organic matter profile. For this, we selected isolated individual trees, such as those found in the Guanica dwarf forest, that provide the ideal setting to study tree species in complete isolation [24].

2. Methodology

2.1 Study site

The study was conducted in the Guánica dry forest, a Biosphere Reserve established in 80's (for the specific location, please see **Figure 1** in [31]) that has been disturbed by humans (urbanization, selective logging and charcoal pits) [32, 33] and hurricanes such as Hurricane Maria in 2017, but at the time of the study, the eyes of less severe hurricanes, such as Hortense (1995) and Georges (1998), had crossed the forest [34]. Average temperature is 25.1°C and precipitation is 860 mm (range 288–1348 mm) but the monthly distribution of the rain is highly erratic [35]. In average, there are 6 dry months (3–8 mo.) that occur as one major dry period from December to April and a minor one from June to August [36]. The forest presents several vegetation associations [37–39], the coastal vegetation is an open forest that occurs on limestone hills with rock depressions that accumulate highly fertile organic matter [33] where trees grow dwarf and isolated from neighbor trees by exposed rock [39]. Here five representative tree species were selected: *Coccoloba uvifera* (evergreen) and *Conocarpus erectus* (evergreen) both present near the cliff shore, and *Tabebuia heterophylla* (facultative deciduous), *Pisonia albida* (obligate deciduous) and *Ficus citrifolia* (deciduous) which are present from the coast to the upper ridges in the forest. Ten trees belonging to each species were selected for a total of 50 trees which represent the sampling units. Under each tree, organic matter samples were collected and divided into fractions showing progressive stages of decomposition that were loose litter (recently fallen leaves), old litter (fragmented leaves) and humus. During the study (from September 2004 to November 2005), the structure of each tree was measured once while microclimate was measured and arthropods were collected under each tree for one year encompassing wet and dry periods, the wet period received unusual high precipitation. The data for each fraction was kept separately.

2.2 Mesoclimate

Mesoclimate was characterized by using data obtained from the NRCS [40] and RAWS sites [41]. The first site is a SCAN (Soil Climate Analysis Network) weather station that is located in an open area in the forest and up 165 m from the coastal plateau while the second site is a RAWS (Remote Automated Weather Stations) weather station located near the coast but the data for the study period is too fragmentary. Temperature and precipitation were taken from the SCAN site, except for October 2005, that is missing and was then was calculated by difference between the previous and next month. This datum was corroborated with the RAWS site. This information is presented in **Figure 1**.

2.3 Microclimate

Microclimate data was obtained from data loggers. For each of the study species, three (out of five) trees were selected to ensure trees were interspersed within the study area and therefore represented the local variation. Data loggers were placed under each tree as follows: one HOBO temperature/humidity data logger was placed at 25 cm above ground, and will be referred as understory temperature throughout the chapter, one TidBit temperature data logger was placed in the old litter fraction (~2.5 cm depth) and one TidBit temperature data logger was placed in the humus fraction (~4.5 cm depth). Data loggers collected temperature data every hour from September 2004 to November 2005. Although the HOBO data logger collected temperature and humidity, only temperature is presented because the humidity sensor shorted out as it was exposed to a salty environment and produced unreliable data.

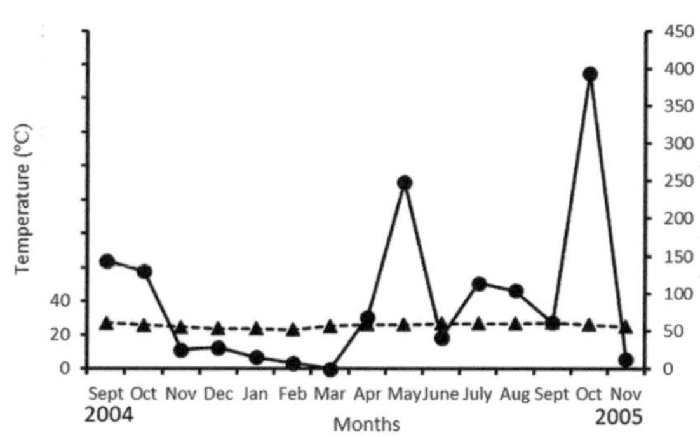

Figure 1.
Diagram showing the relation between average monthly temperature (dashed line) and total monthly precipitation (solid line) in the Guánica forest during the study period. When the solid line is below the dashed one, this indicates water deficit (dry period), when the solid line is above the dashed one this indicates water availability (wet period), and when precipitation surpasses 100 mm, this indicates water surplus. Please note that months begin in September 2004.

2.4 Tree species characterization

Each tree was characterized by measuring tree height, canopy area, organic matter dry mass and depth. Tree height was measured as the distance between the ground and the highest canopy point in vertical orientation, and canopy length and width were measured in horizontal orientation. Canopy area was calculated from length and width. Organic matter dry mass and fraction depth were measured separately for each tree and fraction in November 2004, February, April, June, September and November 2005.

2.5 Arthropods

Arthropod collections were performed under each tree on November 2004, February, April, June, and September 2005. During each sampling, one 10 cm x 10 cm sample/tree/species was collected, and the sample was separated into three fractions: loose litter, old litter and humus. Each fraction was kept separately and placed in a Berlese funnel for one week for arthropod extraction using light [42]. This sampling design gave 5 species x 10 trees x 3 fractions x 5 samplings = 750 samples. Collected arthropods were taxonomically identified to the lowest category possible, either class, subclass, order or suborder, and classified as adult or immature, and assigned to a morphospecies. Collembolans were not assigned to morphotypes since variation in the morphology can only be seen in mounted slides and by a specialist. The abundance of each morphospecies was recorded and standardized to number of individuals per square meter. Morphospecies composition was used as a surrogate for species composition.

2.6 Statistical analyses

Anovas were used to evaluate differences in tree structure and organic matter among plant species. Also, a two-way Anova was used to establish the effect of plant species and fraction on the modulation of temperature. Modulation was estimated as the maximum, minimum and daily range in temperature. To assess the effect of tree structure on temperature modulation we used spearman rank correlations. To assess the effect of time (sampling dates), plant species and fraction on the abundance and richness of arthropods three-way AOV were used. The abundance of arthropod morphotypes was used in a Multi-Response Permutation Procedure (MRPP) to evaluate the effect of time, plant species and fraction on the species composition of adult arthropods. MRPP is a non-parametric test that calculates a distance matrix of average observed distance within predefined groups and compares it to an average distance expected by chance. Within group distance is used to calculate A, a homogeneity parameter of within group variability that ranges between zero and one. In community ecology, values close to zero are common and suggest a heterogenous community that can still be different from other communities [24, 43–45].

3. Results and discussion

3.1 Mesoclimate

During the study, temperature decreased from September to February, and then it increased again (**Figure 1**) and total precipitation was 1575 mm, and was distributed as follows: 480 mm from September to October 2004 producing wet conditions at the end of 2004, 120 mm between November 2004 and April 2005 producing dry conditions at the beginning of 2005, and 975 mm between May and October 2005 producing wet conditions in the second half of 2005 that were interrupted by a short water deficit in June 2005 (**Figure 1**). In general, during the study, the pattern was similar to the one historically established for the Guanica forest [36, 46] with some variations. Historically, the dry season runs from January to July interrupted by a small pulse in precipitation in May, and then follows a wet season from August to December with water surplus during September. This study encompassed two wet periods and one dry period, there was water surplus in September and October 2004, followed by dry months up to April 2005 indicating 6 consecutive months of water deficit. Then, there were wet conditions after April 2005 with three pulses of water surplus in May, July and October 2005. This indicates that during the study, precipitation was atypical because dry conditions lasted 6 consecutive months while Lugo et al. [36] found that, for this forest, dry conditions usually last 3–4 consecutive months. In addition, historically the wet season usually presents one pulse of water surplus but during wet 2005 there were three pulses in precipitation. These data confirm that precipitation in Guánica is highly erratic [35] and show that the 2004–2005 months encompassed in this study represent extreme dry conditions followed by extreme wet conditions.

3.2 Microclimate

3.2.1 Variation among tree species

There was a significant effect of plant species and fraction on maximum, minimum and temperature range (**Table 1**, **Figure 2**). Understory maxima temperatures ranged between 35.9°C and 34.9°C and followed the pattern *Conocarpus* > *Pisonia* > *Tabebuia*, *Ficus*, *Coccoloba* while minima followed the pattern *Tabebuia* ≥ *Coccoloba*, *Ficus*, *Pisonia* ≥ *Conocarpus*. Therefore, in the understory, temperature daily range was largest in *Conocarpus* intermediate in *Pisonia* and smallest in *Ficus*, *Tabebuia*, *Coccoloba*. In the litter, maxima temperatures ranged between 37.8°C and 30.2°C, and followed the pattern *Conocarpus* > *Ficus*, *Pisonia* > *Pisonia*, *Tabebuia* > *Coccoloba*. Minima temperatures were *Pisonia* ≥ *Ficus* ≥ *Conocarpus* ≥ *Tabebuia* ≥ *Coccoloba*, therefore temperature daily range in the litter was largest in *Conocarpus*, intermediate in *Ficus* followed by *Pisonia* and *Tabebuia*, and smallest in *Coccoloba*. In the humus, maxima ranged between 31.7°C and 28.2°C and the pattern was *Conocarpus*, *Ficus* > *Pisonia*, *Tabebuia* > *Coccoloba* and in the humus between. Minima were *Conocarpus* > *Pisonia* > *Tabebuia* > *Ficus* and *Coccoloba*. Therefore, humus

Factor	DF	Maxima		Minima		Range	
		F	p	F	p	F	p
Species	4	576	<0.001	26	<0.001	447	<0.001
Fraction	2	2767	<0.001	3138	<0.001	5624	<0.001
Species x fraction	8	149	<0.001	27	<0.001	142	<0.001
Error	18,995						
Total	19,009						

Table 1.
Effect of plant species and fraction (location: understory, litter, humus) on the magnitude of temperature fluctuations during the study period. Anovas were run using daily maximum, minimum and temperature range. There were 435 days (from September 8, 2004 to November 16, 2005).

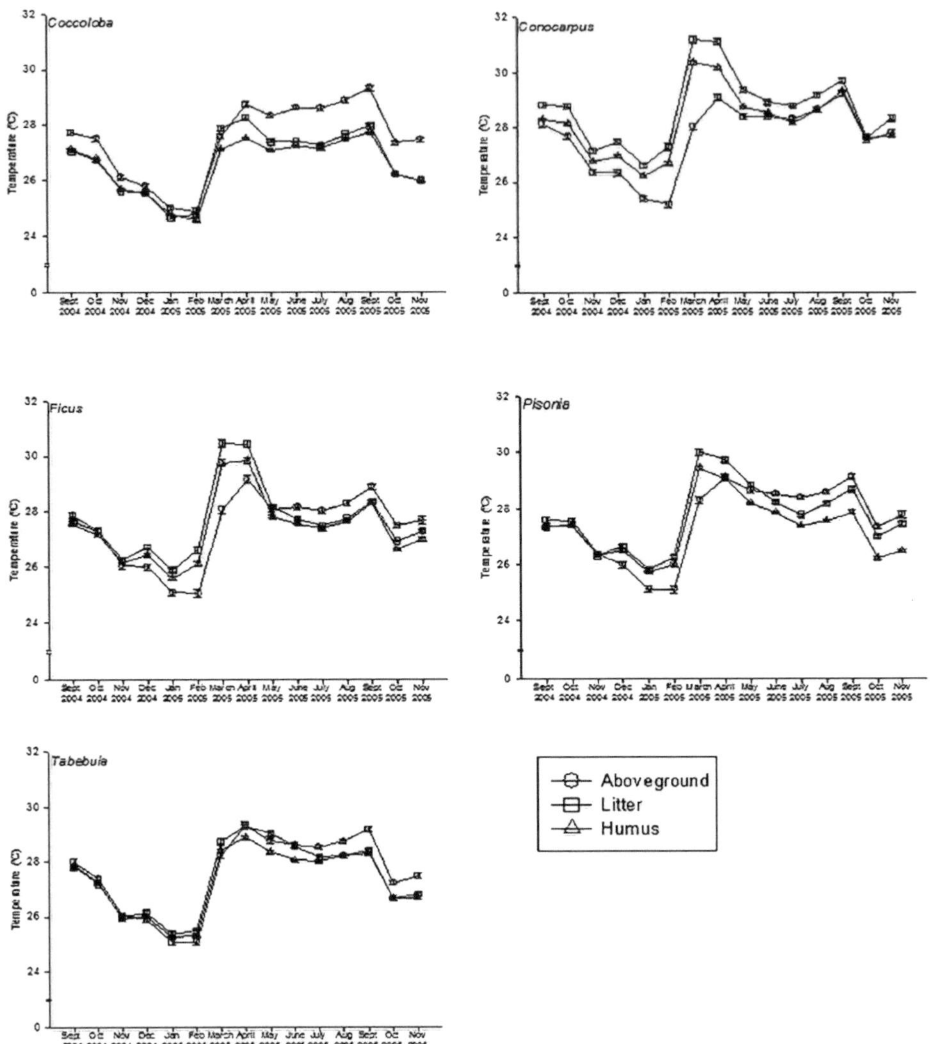

Figure 2.
Mean monthly temperature (±s.e.) in the understory, litter and humus below the five tree species. (understory: ~25 cm understory, Litter: inside the old litter fraction ~2.5 cm depth, Humus: inside the humus fraction ~4.5 cm depth). Symbols represent mean values and whiskers represent standard error.

temperature daily range showed the pattern *Ficus* > *Conocarpus* > *Tabebuia* > *Pisonia* > *Coccoloba*.

Throughfall and stem flow are affected by tree size and differentially moisten soil under canopies [47] while surrounding rock similarly warms the OM underneath trees [48]. In this study temperature range was largest in the litter suggesting that the buffer capacity of humidity was absent in litter (upper OM) and present in humus (deeper OM). On the other hand, among the five species, *Coccoloba* has the largest leaves and *Conocarpus* the smallest ones, and canopy openness is intermediate in *Tabebuia, Ficus* and *Pisonia* [49] suggesting that tree characteristics play a role in modulating temperature fluctuations.

3.2.2 Variations among seasons

As expected, we found that the pattern of temperature fluctuations under all species (**Figure 2**) was similar to the pattern of mesoclimate temperature fluctuations (**Figure 1**) (Kolmogorov Smirnov, $p = 1.00$), e.g. lower temperatures in January and February in comparison to the remaining months. Also, understory air temperature under the five species was consistently higher than temperature from the mesoclimate site by 2.13°C on average. This is due to differences in altitude, the mesoclimate SCAN site was located 165 masl while trees were located at 0 masl, and since temperature decreases by ~1°C for every 100 m in altitude, therefore this resulted in the consistent higher temperatures under the trees. Across months, understory, litter and humus temperature fluctuations were similar for all species, and the largest temperature increase occurred from February to March, an increase of 2.98°C (**Figure 2**). Our data is consistent with the historical pattern described for this forest [36] and with other studies that established that different vegetation associations within a geographic area follow similar fluctuations [50] because fluctuations at the regional scale drive the seasonal pattern at smaller scales, i.e. tree understory temperature.

In all species, through time, daily range significantly varied in all fractions and was larger from December 2004 to March 2005 (**Figure 3**). In *Coccoloba*, *Pisonia* and *Tabebuia* daily variation was largest in the understory and progressively smaller in litter and humus, while in *Conocarpus* and *Ficus* daily variation was largest in the litter particularly in March 2005. During the first quarter of the year, Puerto Rico is affected by cold temperate fronts that decrease average minimum temperatures from December to March [51] but do not affect maxima temperatures producing larger daily variations for all species when compared to remaining months. The largest variation in daily temperature occurred in March 2005 (**Figure 3**), when there was no precipitation (**Figure 1**).

Therefore, in Guanica, months of water deficit (**Figure 1**) coincide with cool months suggesting that mesoclimate drives the decrease in rainfall and minima temperatures while plant species modulate maxima temperatures. In addition, temperature fluctuations in the soil are buffered by soil moisture [52] and isothermal karst rocks warm OM [48] therefore water deficit [53] in combination with an increase in temperature better explain the largest daily variations found in March 2005.

3.3 Tree species characterization

On average, *Pisonia* trees were 2.0 m height, *Coccoloba*'s were 1.8 m and *Ficus*' 1.8 m, and significantly taller than *Tabebuia*'s and *Conocarpus*' that were 1.2 m and 1.1 m respectively (**Figure 4**). Canopies were significantly larger in *Coccoloba*, 155 m^2, than in the other species (**Figure 4**) therefore *Coccoloba* trees were tall and large, *Ficus* and *Pisonia*'s were tall and small, and *Conocarpus* and *Tabebuia*'s were short and small. Litter weight varied among species and through time (**Figure 5**), there was a trend for *Coccoloba* to have the largest litter mass and *Tabebuia* to have the lowest. Litter depth tended to be higher in *Pisonia* particularly at the beginning of the dry season, i.e. January 2005 (**Figure 5**). On the other hand, humus weight and depth were similar among species and months suggesting that although new litter is produced, there is an effect of past accumulation. In Guánica, the ground surface is irregular presenting small depressions of varying sizes [33], then variations in humus may be due to changes in size of the exact sampling points and not because of changes in soil organic matter per se.

Across months, litter fractions varied significantly (**Figure 5**). For example, litter depth under *Coccoloba* was highest in November 2004 and lowest in June 2005, while it was highest in *Pisonia* during the dry months possibly because this species is deciduous and drops the leaves during the dry season [54, 55]. *Ficus* tended to have large litter depth in the wet months; as a facultative deciduous

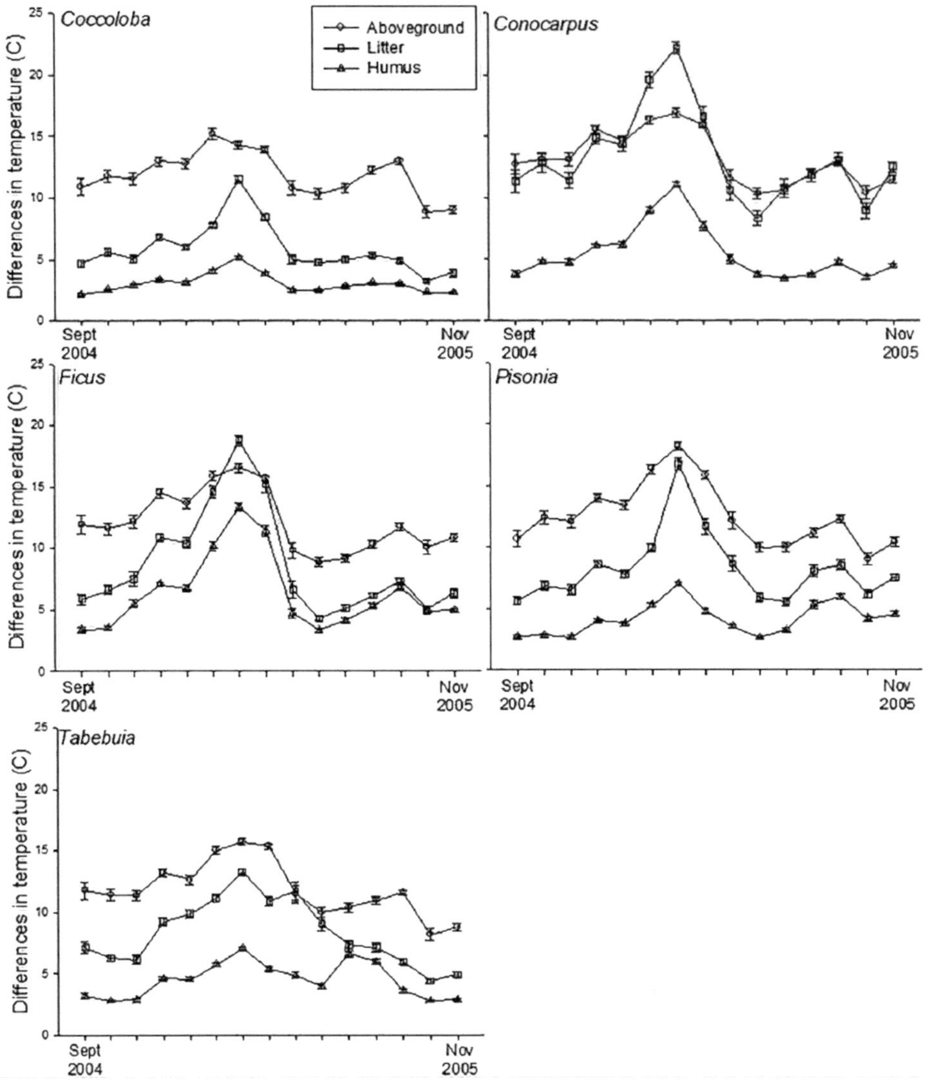

Figure 3.
Mean daily temperature range per month below the five tree species and for each of the three depths. Daily range was calculated as Tmax – Tmin/day. Symbols represent mean values and whiskers represent standard error.

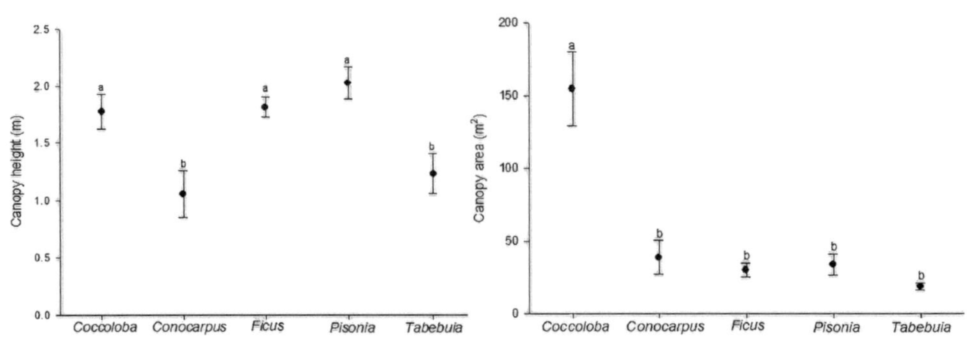

Figure 4.
Average canopy height (m) and average canopy area (m^2) for the five species. Dots represent means, whiskers represent standard error, and lower-case letters indicate significant differences.

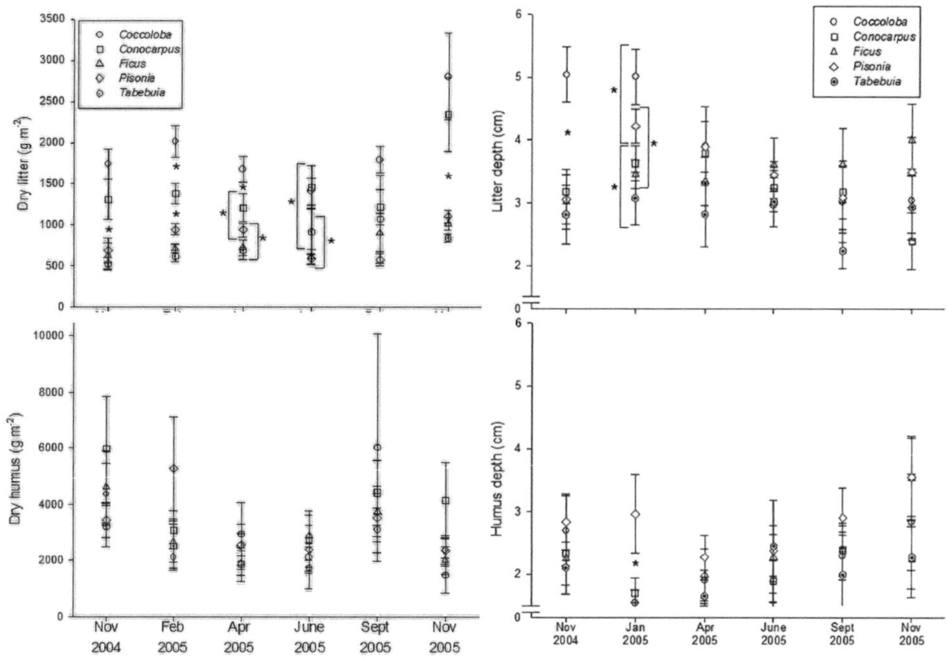

Figure 5.
Average litter and humus dry weight (g m^{-2}) and depth (cm) for the five tree species and in each of the six sampling dates. Symbols indicate the mean and whiskers indicate standard error. Asterisks indicate significant differences among plant species within censuses.

species, it drops the leaves during the dry season [54, 55] but these leaves had a slower decomposition possibly in relation to milky sap that deters feeding organisms than did *Pisonia*. *Tabebuia* and *Conocarpus* consistently had small litter depth possibly due to low production of litter fall. Our results are consistent with Lugo et al. [36] who found that litter production increased during the dry season because of deciduous species and because decomposition was arrested (due to water deficit). Humus is a stable fraction of soil organic matter, and loose and old litters are the dynamic fractions, then any change in soil organic matter depth can be attributed to changes in litter.

3.4 Tree species and microclimate

In summary, we have shown that microclimate fluctuations follow the seasonal pattern of mesoclimate, that maxima and temperature range, tree height, canopy area and litter are different among plant species. Different tree species characteristics correlated significantly with temperature, and these correlations varied seasonally (**Table 2**). In general, understory air temperature correlated with tree characteristics in the dry season but not in the wet season, while litter and humus temperatures correlated with tree characteristics in both seasons and with litter mass only in the wet season. Canopy size correlated significantly with understory maximum temperature in February 2005 when deciduous species drop the leaves [37] suggesting that more open canopies allowed for higher maximum temperatures in the dry season. On the other hand, litter mass and depth correlated with litter temperature suggesting that the quantity of organic matter and its water holding capacity is important for buffering litter temperature during the wet season.

Canopy area significantly correlated with temperature, for example maximum temperature was highest in *Conocarpus*, a species with small canopies while species

	February 2005			September 2005		
	Maxima	**Minima**	**Range**	**Maxima**	**Minima**	**Range**
Understory	• Canopy size** • Tree volume***		• Canopy size** • Tree volume***			
Litter	• Canopy size*** • Tree volume*** • Tree height*	• Tree height*** • Litter depth***	• Canopy size*** • Tree height*** • Tree volume*** • Litter depth***	• Canopy size*** • Tree volume*** • Litter depth*** • Litter mass***	• Tree volume*** • Litter mass***	• Canopy size*** • Tree volume*** • Litter mass***
Humus	• Canopy size*** • Tree volume***		• Canopy size*** • Tree volume***	• Canopy size*** • Tree height*** • Tree volume*** • Litter depth*** • Litter mass***	• Canopy size*** • Tree volume*** • Litter depth*** • Litter mass***	• Canopy size*** • Tree height*** • Tree volume*** • Litter mass***

*Only significant correlations are shown, and asterisks next to the variable indicate level of significance: *: $p < 0.05$, **: $p < 0.01$, ***:$p < 0.001$. February and September 2005 are shown as representatives of dry and wet periods.*

Table 2.
Spearman rank order correlations among maximum, minimum and range in temperature and tree characteristics (height, size, volume, organic matter depth and volume).

with similar canopy areas such as *Ficus*, *Pisonia* and *Tabebuia* had lower maximum temperatures, and *Coccoloba* had taller trees with larger canopy areas which possibly resulted in less understory heating and lower temperatures. Leaf area and leaf orientation result in light differentially entering plant canopies producing different degrees of heating of understory air [1, 52]. Therefore, our data suggest that the size of the tree (canopy and height – volume) influence the microclimate created underneath it but not in a linear additive way, rather tree characteristics interact and influence microclimate underneath the tree in complex ways, suggesting that the tree is an integrate that modulates the understory microclimate.

3.5 Arthropods

3.5.1 Litter fractions

A total of 8702 arthropods representing 22 orders and 301 morphotypes were collected. Arthropod abundance and richness were significantly different among fractions (**Table 3**). Overall *Tabebuia* and *Ficus* had similarly higher abundance, followed by *Pisonia* and *Coccoloba*, and abundance was lowest in *Conocarpus*. In general, there was a trend for lower arthropod abundance in loose litter and highest in humus except in *Coccoloba* where old litter had higher abundance than the other two fractions (**Figure 6**). Overall, richness was highest in *Ficus* and *Pisonia*, followed by *Tabebuia*, and lowest in *Coccoloba* and *Conocarpus*, but among fractions the trend for lower richness in loose litter and higher in humus was similar as for abundance (**Figure 6**). These data suggest that for arthropod abundance and

		Abundance		Richness	
Factor	**DF**	**F**	**p**	**F**	**p**
Date	4	9.36	<0.001	41.68	<0.001
Species	4	1.65	0.16	3.39	0.01
Fraction	2	40.83	<0.001	66.13	<0.001
Date x species	16	0.90	0.57	1.52	0.09
Date x fraction	8	2.92	0.00	7.23	<0.001
Species x fraction	8	2.59	0.01	1.93	0.05
Date x species x fraction	32	0.79	0.79	0.93	0.58
Residual	675				
Total	749				

Table 3.
Three-way analysis of variance for the effect of date, species and fraction on the abundance and richness of arthropods.

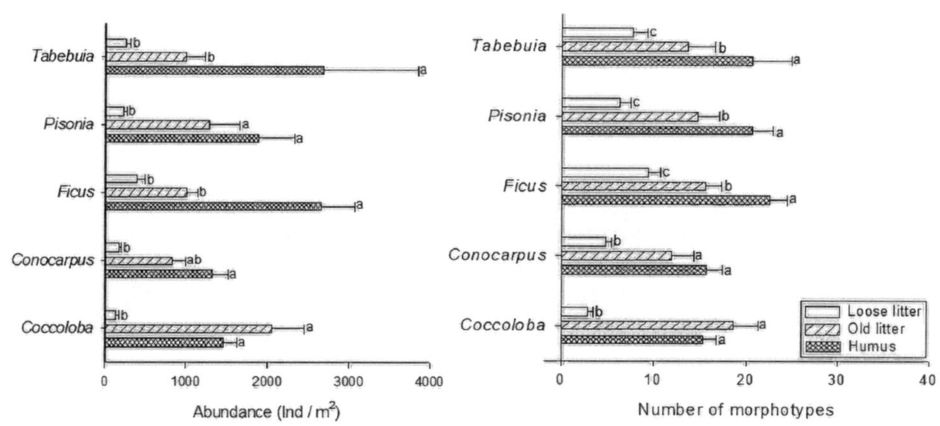

Figure 6.
Average arthropod abundance (±s.d.) (ind/m2) and average number of morphotypes (±s.e.) in each fraction for the five study species. Lower case letters indicate significant differences among fractions within species.

richness, the decomposition stage of the organic matter (paralleled in the vertical profile: loose, old, humus) is a determining factor as important as plant species identity [24]. Furthermore, the decomposition stage is advanced in humus where the biological activity is high and thus there are high available resources (e.g. higher nutrient availability in humus than in litter fractions, [11]) suggesting a higher abundance of decomposer microbes that, in the trophic food web theory would result in higher abundance of arthropods.

Hansen [12] found that large litter dwellers were positively associated with particular litter substrates, Berg et al. [22] similarly found that litter in different decomposition stages had sets of associated arthropods, and Prinzing et al. [56] found that even sexes within a mite species differentially used organic matter strata. At the same time, Hansen [12] proposed that a higher abundance and richness of arthropods would be found in strata with higher diversity of resources (i.e. litter). Our results support that specific groups are associated to specific litter strata, being more abundant where more resources are found, but not in the litter, instead in the humus fraction. Kardol et al. [57] suggest that plant–soil feedbacks are mediated by microbes; specifically, they found that soil microbial pathogens affected the development of the future plant community because soils have a biotic legacy of past vegetation cover. Similarly, soils may have a biotic legacy of beneficial microbes that, as Wardle states, maximizes decomposition, and can have a positive historical feedback on the soil community. In addition, Bezemer et al. [58] found that plant–soil feedbacks are dependent on plant species. In our case, these data suggest that trees in the coastal plateau have been there for long enough to result in a biotic legacy that, under certain plant species favors some arthropods, while under other plant species other types of arthropods are favored.

3.5.2 Variations through time

Arthropod abundance and richness were significantly different among sampling dates (**Table 3**). Through time, arthropod abundance varied with a trend for a decreased abundance in the dry period (April 2005) and higher abundance in the wet period (**Figure 7**). In all species and all samplings, arthropod abundance was consistently low in loose litter while variations in old litter and humus explained significant differences. For example, in *Coccoloba*, *Ficus* and *Tabebuia* abundance peaked at the onset of the wet period (June 2005) while in *Conocarpus* and *Pisonia* peaked abundance was observed later in the wet period (September 2005). Similarly, for all plant species, arthropod richness was significantly low in the dry period and higher in the wet period (**Figure 8**). Also, in *Coccoloba*, *Ficus* and *Tabebuia* peaked richness occurred at the onset of the wet period (June 2005) while in *Conocarpus* and *Pisonia* it was later in the wet season (September 2005).

During the study, precipitation varied producing a wet 2004 period, a dry 2005 period and a wet 2005 period (**Figure 1**). Herrera [59] found that the abundance of soil fauna fluctuated seasonally influenced by the precipitation regime. Specifically, groups like ants and beetles were present during year-round but isopods and diplopods were more abundant at the end of the wet season while chilopods were present all the year but during the dry season they migrated vertically and thus were completely absent in the upper litter. On the other hand, Prather et al. [27] found that high soil moisture promoted arthropods while high temperature decreased arthropods. Our results show that arthropod abundance and richness follow a seasonal pattern tight to precipitation, and that arthropods tightly linked to litter are able to overcome drought although in such low abundance that does not support upper trophic levels [24].

Arthropod species composition was dynamic through time as it varied among plant species (**Figure 9**). Under wet conditions such as in November 2004, arthropod species composition was similar between *Coccoloba* and *Conocarpus* and among *Tabebuia, Ficus* and *Pisonia* who shared unique sets of associated arthropods. Similarly, in June and September 2005, a similar pattern was observed such that species composition was similar between *Coccoloba* and *Conocarpus*, and among *Ficus*, *Pisonia* and *Tabebuia.* On the other hand, at the onset of the dry period (February 2005) when water shortage peaked, differences among plant species were lost resulting in similar arthropod species composition among species (**Figure 9**). Therefore, in general, when water is available, arthropods formed two distinctive communities under different groups of plant species, while when water deficit happened, plant species shared a common set of arthropods.

From these data, we conclude that two major arthropod communities were formed, one with arthropods common to *Coccoloba* and *Conocarpus*, and one with arthropods common to *Ficus*, *Pisonia* and *Tabebuia*. The local distribution of trees shows that *Coccoloba* and *Conocarpus* are closer to the shore than the other three species, suggesting that plants near the coast share arthropods communities as do

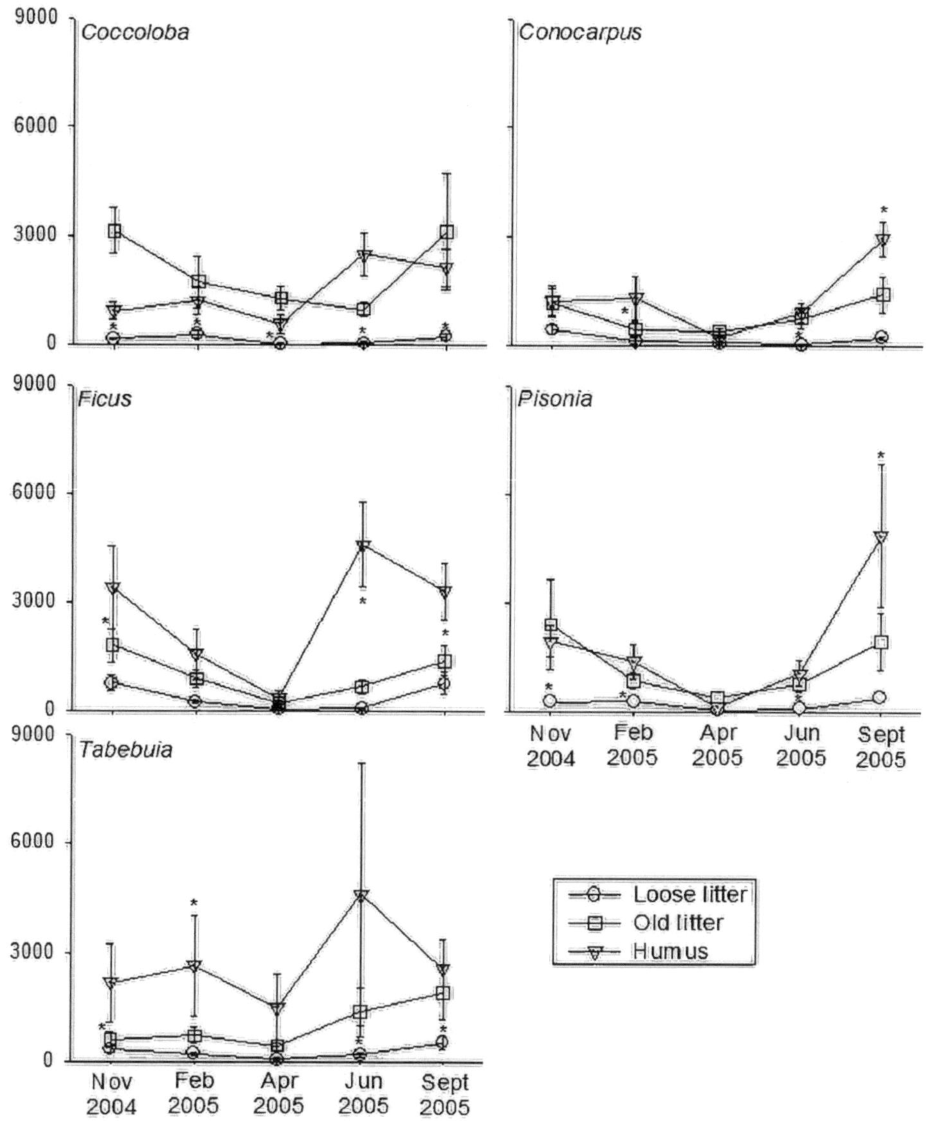

Figure 7.
Mean arthropod abundance (±s.d.) (ind/m²) in each sampling date and in each fraction for the five study species. Asterisks indicate significant differences among fractions within species for the corresponding sampling date.

plants farther from the shore. Nevertheless, when extreme conditions happen, such as dry cool months in February 2005 (**Figure 1**), the arthropod community homogenizes because arthropod abundance was low and species composition becomes similar.

Both, arthropod abundance and richness (**Figures 7** and **8**) decreased at the onset of the dry period (February 2005) and were at a minimum in the middle of the period (April 2005), on the other hand species composition was similar at the onset but different in the middle of the dry season (**Figure 9**). Furthermore, litter depth was high at the onset of the dry season (**Figure 5**) because of new litter produced by plants as a response of water deficit (**Figure 1**). These results suggest that during the dry season, conditions are such that plants drop leaves and belowground arthropods associated to plants represent a shared subgroup common to all

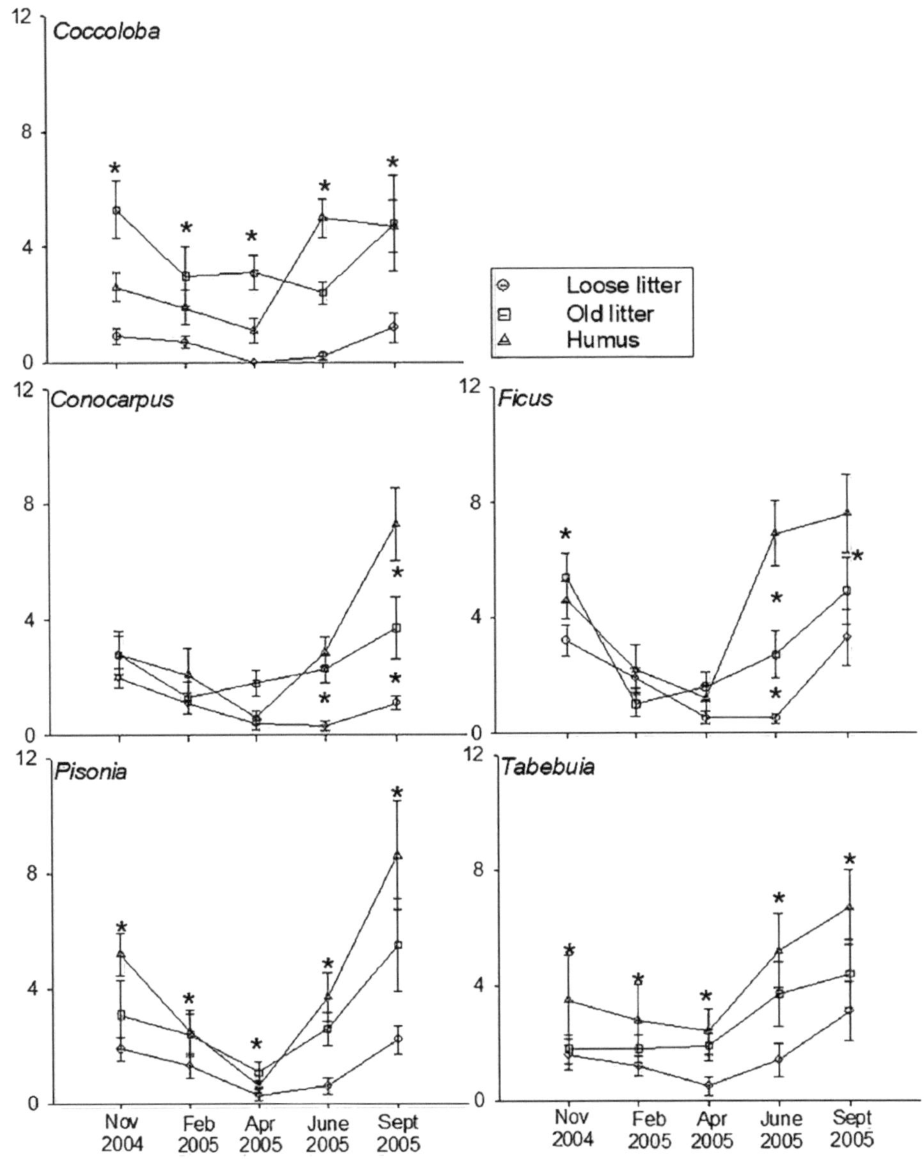

Figure 8.
Average number of morphotypes in each sampling date and in each fraction for the five study species. Asterisks indicate significant differences among fractions within species for the corresponding sampling date.

plant species, while further into the dry season, arthropod abundance and richness are at the lowest such that only arthropods specific to plant species are present. Abundance and richness are not linearly related, nevertheless an increase in abundance results in more species, for example Prather et al. [27] found that for every 16 individuals a new species appears. From these data we can expect that a decrease in abundance results in loss of species, therefore, in our study the decrease in abundance at the beginning of the dry season may have resulted in a decrease in species such that only stress-tolerant arthropod species [27] were present in all plant species producing a homogenization of the arthropod community. A further decrease in abundance led to the minimum found in mid-dry season which might have produced further species loss so that only species adapted to the specific plant microhabitat remained resulting in a differentiation in arthropod community.

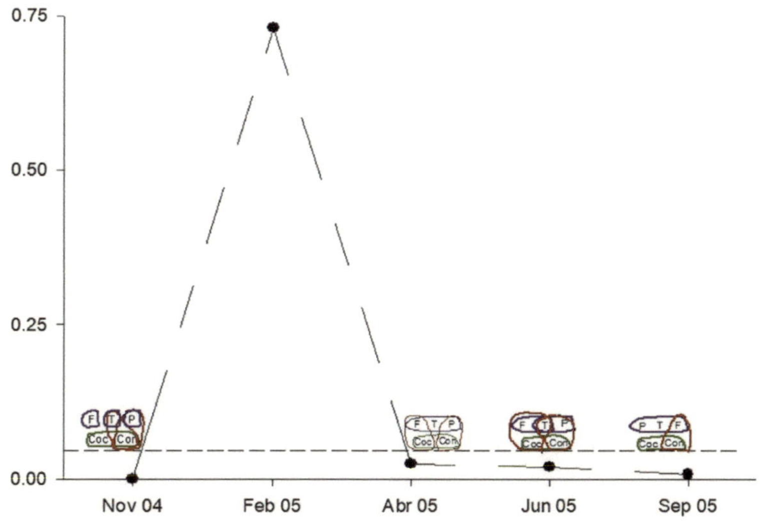

Figure 9.
p-value results from MRPP (Multi-response permutation procedure) tests. MRPP evaluated whether there were significant differences in arthropod species composition among plant species in each sampling. The y-axis shows p-values, initials indicate plant species as follows Coc *for Coccoloba,* Con *for Conocarpus,* F *for Ficus,* P *for Pisonia and* T *for Tabebuia. The short-dashed horizontal line indicates the critical value p = 0.05. Markers below the dashed line show significant differences among species. Significant differences are shown by circled color-coded initials.*

Species	Nov 04 – Feb 05	Feb – Apr 05	Apr – June 05	June – Sept 05
Coccoloba	81(±8)**	96(±5)**	77(±7)	73(±9)*
Conocarpus	94(±3)*	94(±7)*	82(±9)	86(±9)*
Ficus	92(±5)*	91(±4)	91(±3)**	79(±6)
Pisonia	92(±4)*	94(±6)	89(±6)	89(±5)***
Tabebuia	94(±4)	94(±6)	88(±8)	85(±5)

*Percent dissimilarity was calculated as 100 * dissimilarity. Dissimilarity was calculated using the Sorensen index (1-2 W/(A + B)). W indicates the sum of abundances of common species between sampling units, and A and B indicate the sum of abundances of all species in each sampling unit. Asterisk indicates significant differences among consecutive sampling dates (*:p < 0.05, **:p < 0.001, ***:p < 0.0001).*

Table 4.
Dissimilarity (%) of arthropod species composition (mean ± s.d.) associated to specific plant species among sampling dates.

Within each plant species, arthropod species composition significantly varied through time (**Table 4**). In *Coccoloba* and *Conocarpus*, species composition significantly changed between consecutive samplings except April to June 2005. These data suggest that for these two plants there is a high turnover of arthropod species, and that the arthropods present in April were a subset of the June arthropods. When comparing April and June 2005, in *Coccoloba*, abundance and species composition were similar and richness increased (**Table 4**, **Figures 7** and **8**) suggesting that at the onset of the wet season the majority of arthropods belong to the same species despite an increase in richness. In *Conocarpus*, when comparing April and June, arthropod abundance (**Figure 8**), richness (**Figure 9**) and species composition (**Table 4**) were similar suggesting that the transition between dry and wet months is dominated by a common set of arthropods.

In *Ficus*, the arthropod composition was similar through time except from November to February and from April to June (**Table 4**), suggesting that the major shifts in arthropod species occur between seasons (end wet period: November to February, end dry period: April to June). On the other hand, arthropod abundance and richness decreased to a minimum in the middle of the dry season (April 2005) (**Figures 8** and **9**) when litter increased (**Figure** 5) because this is a deciduous plant. These data suggest that arthropods present in *Ficus* respond to changing conditions, such as new litter or a change in precipitation.

In *Pisonia*, a similar pattern as in *Ficus* was observed but later in the wet period (June to September), arthropod abundance and richness increased in the wet period (**Figures 6** and 7), and species composition was significantly different (**Table 4**). In addition, *Pisonia* is an obligate deciduous species that produces new litter in the dry season (**Figure 5**). These data suggest a possible time lag tied to increasing precipitation as arthropods present in *Pisonia* increased and litter depth decreased (**Figure 5**) probably explained by decomposition of new leaves and more arthropods at the onset of the wet season.

In *Tabebuia*, arthropod species composition and abundance were similar through time (**Figure** 7 and **Table 4**) while richness (**Figure 8**) significantly varied suggesting that there is a bulk of arthropods associated to this plant species that is present yearlong although some arthropod species vary through time but not enough to produce a different composition of arthropods.

For most plant species, these data show a shared pattern for a shift in arthropod communities between wet and dry seasons (November 2004 to February 2005), then arthropods in *Cocoloba* and *Conocarpus* continued to change as the dry season strengthened while arthropods in *Ficus* and *Pisonia* (deciduous plants with more open canopies by February 2005) changed with an increase in rainfall that triggers biological activity. *Tabebuia* was the only plant to show a consistent group of associated arthropods.

3.5.3 Variations across fractions

Abundance and richness of arthropods were significantly affected by sampling time and fraction (interaction term Date x Fraction in **Table 3**), suggesting that through time there is a dynamic movement of arthropods among fractions (**Figures** 7 and **8**). In addition, for each plant species the composition of arthropods varied across the vertical organic matter profile (**Figure 10**).

In *Coccoloba*, there was a general trend for high turnover of arthropods which were more abundant and species composition was different among fractions in the wet season and similar in the dry season when abundance was lowest (**Figures 8–10**). In wet months, loose litter had 9–12 and old litter had 27–30 morphotypes and shared 1–4, while humus had 14–25 from which 12–16 occurred in old. Loose and humus fractions had 1 to none arthropods in common (Appendix I). These data show a unique arthropod community in loose litter mainly due to adult Diptera (unique to this fraction) that possibly pupated and emerged as adults at the moment of collection (e.g. in November: G-157, G-158, G-159, and in September: G-182, G-123, G-191, G-209) (Appendix I). On the other hand, humus species composition was unique in June due to a dominance of detritivore and predator species exclusive to this fraction, such as G-207, G-007, G-147, G-229 and G-263 (Appendix I). These data suggest that the beginning of the wet season triggers the biological activity in the humus fraction (as suggested by increased arthropod abundance and richness, and unique species composition), and that later in the wet season (September and November samplings) decomposer activity increases (as suggested by the presence of adult Diptera whose larvae feed on decomposing material, e.g. G-157, G-158 and G-159), and that a possible flush in aboveground biomass promotes herbivore activity that pupates in

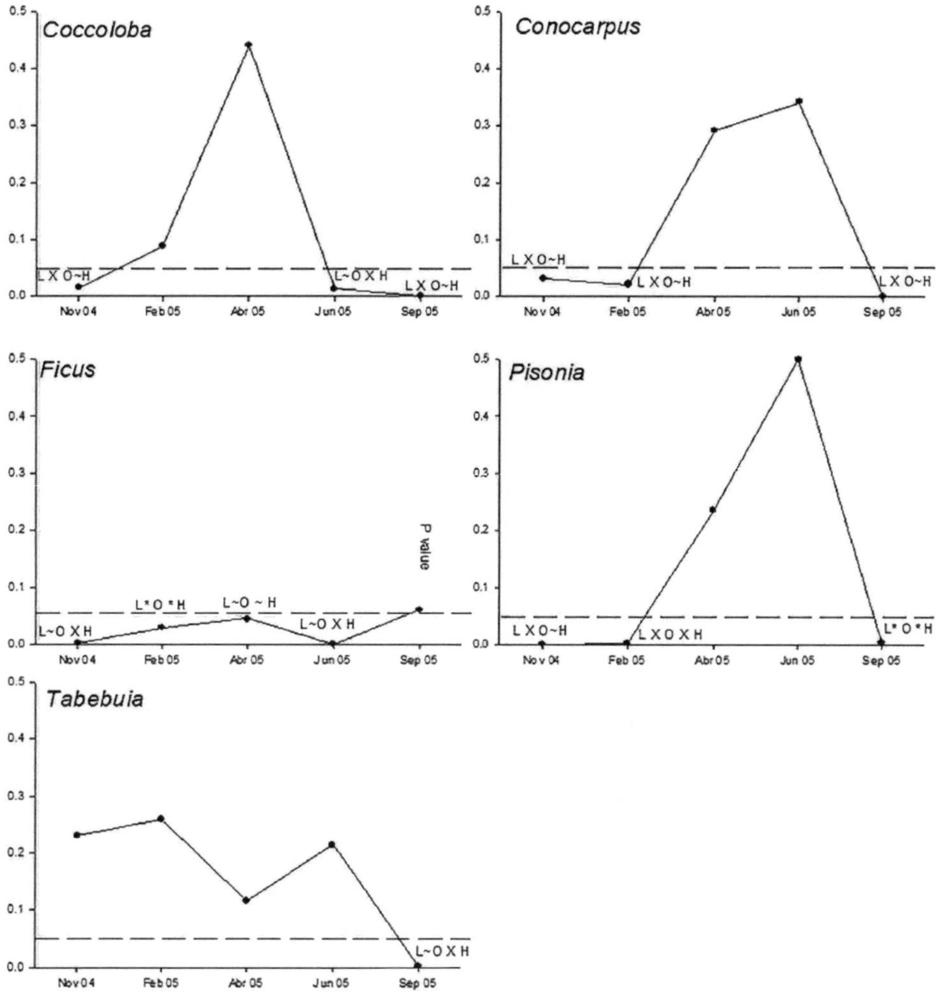

Figure 10.
p-value results from MRPP tests. MRPP evaluated differences among fractions in arthropod species composition, for each plant species at each sampling separately. The y-axis shows p-values, initials indicate fractions as follows L for loose litter, O for old litter and H for humus. The short-dashed horizontal line indicates the critical value p = 0.05. The long-dashed line connects consecutive samplings. Markers below the dashed line show significant differences among fractions. Symbols indicate differences, 'X' indicates significant differences between the specified fractions, ' ~ ' indicates no significant difference, and '' that two fractions are similar but different from the third one.*

the ground (as suggested by the presence of adult Hymenoptera whose larvae are herbivores, e.g. G-168). The similarity in species composition among fractions during the dry period samplings (February and April) suggests that an increase in daily temperature (**Figures** 2 and 3) and drought represent stressors that mix the vertical stratification of the arthropod fauna. For example, in April loose litter had no arthropods, suggesting that remaining arthropods migrated downwards to avoid these extreme conditions.

In *Conocarpus*, also there was a general trend for high turnover of arthropods through time, but there was a consistent pattern for loose litter to have a low abundance arthropod community that is different from old litter and humus (**Figures** 7 and 8). In addition, when abundance was at a minimum in April 2005 (**Figures** 7 and 8), the species composition was similar (**Figure 10**). In November, February and September samplings, loose litter had unique species that were adult Diptera that possibly emerged at the moment of collection, and predators such as

Araneae and Opiliones (i.e. G-134 and G-101) (Appendix I). In April, abundance and richness were lowest especially in loose litter, and species composition was similar among fractions suggesting again that arthropods migrated downwards to avoid extreme conditions such as drought and heat. Although the wet period began in May (**Figure 1**), abundance and richness continued to be low in loose litter and species composition was similar among fractions. In June, loose litter had 3 morphotypes with very low abundance, two were detritivore mites (G-147 and G-100 - Appendix I) suggesting that the increase in precipitation may have promoted detritivore abundance but still not enough to propagate into the loose litter fraction, possibly because *Conocarpus* is a species that grows very near the shore cliff and may receive more salt spray and less effective rainfall so that it may take longer to respond to an increase in water supply.

In *Ficus*, in general, arthropod species composition was different through time and across fractions while abundance was higher in wet months and also in dry months when abundance decreased (**Figures** 7, **8** and **10**). In November and June, humus had a specific set of associated arthropods that were mainly detritivores and predators (i.e. G-007, G-092, G-207, and G-244 - Appendix I). In February and April, species composition was marginally different, possibly due to the presence of adult Diptera in loose and old litter, and large detritivores (such as Psocoptera and Blattodea G-102 and G-228 - Appendix I) in old litter, while in humus there were basically small detritivores and predators (G-207, G-003 and G-220). A possible explanation for the presence of adult Diptera in old litter is that *Ficus* is a facultative deciduous species that progressively drops the leaves in dry conditions so that dipteran larvae that began pupating were then covered by new litter. In September, species composition mixed among fractions possibly due to the shared presence of highly abundant morphotypes such as G-078, G-037, G-105 and G-031 (Appendix I), mainly detritivores, suggesting that in mid wet season, there has been enough detritus production so that detritivores are promoted in the whole organic matter vertical profile resulting in a similar species composition among fractions.

In *Pisonia* also the general trend was for a different species composition when abundance is high in the wet season and it becomes similar when abundance decreases in the dry season (**Figures** 7, **8** and **10**). In November, February and September, abundance and richness were comparatively high, and species composition was different among fractions. In loose litter, the majority of unique morphotypes were adult dipterans (e.g. G-157 and G-139) and adult coleopterans that as larvae are herbivores (e.g. G-146 and G-148) while old litter and humus shared basically abundant detritivore mites (G-037, G-007 and G-087) (Appendix I). In April and June, there was a similar composition of arthropod across fractions, possibly due to the extremely low abundance of arthropods. This may be the result of downwards migration of arthropods and a pulse of new litter because *Pisonia* is an obligate. These data suggest that extreme conditions that occur during the dry season mixed the stratification of arthropods in the vertical organic matter profile.

Tabebuia had a similar arthropod community across fractions through time although abundance was similar and richness increased in the wet season (**Figures** 7, **8** and **10**). The similarity in species composition was possibly due to the shared presence of abundant morphotypes among fractions at each sampling date (e.g. in November G-037, G-003, G-002, G-001, in February G-078, G-188, in April G-037, G-078, in June G-078, G-002, G-004, G-105) (Appendix I). In September, there continued to be abundant shared morphotypes but the appearance of unique detritivore species in humus differentiate the species composition (e.g. as G-262, G-288, G-010, G-269). These data suggest that the arthropod fauna below *Tabebuia* actively moves among fractions and that, at mid wet season, enough detritus has been produced in order to differentiate active arthropods in each fraction.

4. Conclusions

We found abundance and richness to be highest in humus and species composition to be different among fractions suggesting that arthropod communities are segregated among plant species and are further stratified by decomposition stage. Also, abundance and richness were lowest in the dry months when arthropod communities homogenize while in the wet season there was a pattern for arthropods to form two distinctive groups, one formed by arthropods common to *Coccoloba* and *Conocarpus*, and a second group formed by arthropods common to *Ficus*, *Pisonia* and *Tabebuia*. Furthermore, under plant species, the seasonal distribution of arthropods among fractions in different decomposition stages was differentially affected by variations in microclimate and organic matter depth.

We found for 4 out of 5 plant species, that arthropod communities shift at the end of the wet season when water becomes scarce, and again at the onset of next wet season when temperature daily range was largest. In addition, we identified three patterns of arthropod dynamics across fractions.

First *Cocoloba* and *Conocarpus,* two species close to the shore cliff, showed stratification of arthropod communities across fractions except in dry months when arthropods migrate downwards to deep OM. Then at the beginning of the wet season, arthropods are a subgroup of those in the dry season, and as wet seasons continues, abundance increases and arthropods stratify across fractions.

Second *Ficus* and *Pisonia* both showed a seasonal increase in litter and distinct arthropod communities in wet and dry seasons. During the wet season arthropods are abundant and effectively stratify along OM fractions while in the dry season, in general, communities continue to be different across fractions despite a decrease in abundance, and at the onset of the wet season arthropods are a subgroup of those in the dry season. Nevertheless, *Conocarpus* and *Pisonia* present a time lag as compared to *Coccoloba*, in *Conocarpus* possibly related to less effective rain (proximity to the coastal cliff) and in *Pisonia* possibly related to the time that new litter takes to decompose enough to promote arthropods (as an obligate deciduous species).

Third in *Tabebuia,* litter depth was low all yearlong and temperature range was smaller, and we found arthropod species composition to be similar and abundance to vary through time, being always lowest in loose litter, also arthropod composition was similar across fractions suggesting that arthropods actively move among fractions.

Acknowledgements

This research was partially funded by CREST-Center for Applied Tropical Ecology and Conservation of the University of Puerto Rico at Rio Piedras Campus, grant NSF-HRD-0206200 through a fellowship to MFBA. Additional funding and logistic support was provided by the CREST-Center for Applied Tropical Ecology and Conservation of the University of Puerto Rico at Rio Piedras Campus, USDA Forest Service-International Institute of Tropical Forestry and the Guanica Dry Forest, a Biosphere Reserve staff.

Appendix I: Excerpt from appendix in Barberena-Arias (2008). Abundance (ind/m^2) of adult morphotypes in each fraction and species, and for each of the samplings

Date	Order	Morpho	*Coccoloba*			*Conocarpus*			*Ficus*			*Pisonia*			*Tabebuia*		
		Type	L	O	H	L	O	H	L	O	H	L	O	H	L	O	H
Nov. 2004	Pseudosc.	G-001		700	200		100		100	300	200		200	300	200	200	200
Nov. 2004	Acari	G-002	100	1000			100								400	900	1300
Nov. 2004	Acari	G-003		900	400		400	400	100	1100	600	100	900	600	100	700	500
Nov. 2004	Acari	G-004		700	200		100	500	100	600	1000	100	100			400	1000
Nov. 2004	Acari	G-007		2500	300		100				1100		500	700			400
Nov. 2004	Isopoda	G-010		100			100	2800		300		100	100	1100			
Nov. 2004	Acari	G-031		1000													
Nov. 2004	Acari	G-037	100	7100	1900	100	1800	2300	100	400	1800		1500	1300	100	200	1800
Nov. 2004	Acari	G-078		600	700		200	200		300	3800			1600		100	4500
Nov. 2004	Isopoda	G-087					100						400	300		100	100
Nov. 2004	Coleoptera	G-092									200			300			
Nov. 2004	Opiliones	G-101		100	100	100							100			100	100
Nov. 2004	Psocoptera	G-102							300	400			100				
Nov. 2004	Diptera	G-123					100		100	200				100			
Nov. 2004	Araneae	G-134				200			500								
Nov. 2004	Acari	G-100															100
Nov. 2004	Diptera	G-139		100		200			100			100		100			
Nov. 2004	Acari	G-147										100					
Nov. 2004	Diptera	G-157	100			100						100					
Nov. 2004	Diptera	G-182		100				100									
Nov. 2004	Coleoptera	G-146										100					

Date	Order	Morpho	*Coccoloba*			*Conocarpus*			*Ficus*			*Pisonia*			*Tabebuia*		
		Type	L	O	H	L	O	H	L	O	H	L	O	H	L	O	H
Nov. 2004	Coleoptera	G-148										100					
Nov. 2004	Diptera	G-158	100														
Nov. 2004	Diptera	G-159	100														
Feb. 2005	Acari	G-002		500	200					600	100	100	300			600	900
Feb. 2005	Acari	G-003		2300	400						1800	200	100	600		400	1300
Feb. 2005	Acari	G-004	100				100	1000	100		300		900	1100	200		300
Feb. 2005	Acari	G-007		1600	600		200				600		100	500			200
Feb. 2005	Acari	G-037		100	100	100	100	300	100		200			500			500
Feb. 2005	Acari	G-078		100	1700			300		100	100			400	100	100	2900
Feb. 2005	Psocoptera	G-102								100			400			100	
Feb. 2005	Araneae	G-134	200	500		400			300						200		
Nov. 2004	Hymenopt.	G-168	100														
Feb. 2005	Acari	G-188	100	100				200			100				100	100	100
Feb. 2005	Acari	G-207								100	4600						4600
Feb. 2005	Isopoda	G-010												200			100
Feb. 2005	Acari	G-100											100				
Feb. 2005	Diptera	G-139				100			100								100
Feb. 2005	Diptera	G-159								100			200				
Feb. 2005	Diptera	G-191										100			100		
Feb. 2005	Pseudosc.	G-220									200					100	100
Feb. 2005	Araneae	G-229		100			100										

Date	Order	Morpho	*Coccoloba*			*Conocarpus*			*Ficus*			*Pisonia*			*Tabebuia*		
		Type	L	O	H	L	O	H	L	O	H	L	O	H	L	O	H
Feb. 2005	Pseudosc.	G-001						100									
Feb. 2005	Coleoptera	G-092									100						
Feb. 2005	Opiliones	G-101								100							
Feb. 2005	Diptera	G-209							100								
Feb. 2005	Blattodea	G-228															
Apr. 2005	Pseudosc.	G-001		200			100	100									
Apr. 2005	Acari	G-002		200	100					200	100					100	1700
Apr. 2005	Acari	G-003					300				1000						700
Apr. 2005	Acari	G-004		1100	300	100	100			100	100		600			500	9600
Apr. 2005	Acari	G-007								200				100			200
Apr. 2005	Acari	G-037		3400	700		1400	700		100	100		1000		200	300	300
Apr. 2005	Acari	G-078		800	400		100			500	100		300	300	100	1200	800
Apr. 2005	Acari	G-100														100	100
Apr. 2005	Araneae	G-134		100			200			200							
Apr. 2005	Diptera	G-139								100		100					
Apr. 2005	Acari	G-207		1000			100									200	100
Apr. 2005	Diptera	G-123		100													
Apr. 2005	Diptera	G-157									100						
Apr. 2005	Acari	G-188															100
Apr. 2005	Pseudosc.	G-220												100			
Apr. 2005	Blattodea	G-228														100	

Date	Order	Morpho	*Coccoloba*			*Conocarpus*			*Ficus*			*Pisonia*			*Tabebuia*		
		Type	L	O	H	L	O	H	L	O	H	L	O	H	L	O	H
Apr. 2005	Acari	G-244							100								
Jun. 2005	Acari	G-002		500	400		200	200	200	400	1500	100	700	400	200	1200	1800
Jun. 2005	Acari	G-003			1000		100	100		100	2200			100		400	100
Jun. 2005	Acari	G-004		300	1800		400	300		300	500	100	600	600	300	100	900
Jun. 2005	Acari	G-007			1500			100			1900		100			400	500
Jun. 2005	Acari	G-031								300	1300			100		100	
Jun. 2005	Acari	G-037		4900	4100		1100	2200		100	3400		300	500		500	3400
Jun. 2005	Acari	G-078		500	4800		500	1700		300	5700	100	200	1000	100	300	5600
Jun. 2005	Acari	G-105		300	100			200		300	500		200		100	100	400
Jun. 2005	Araneae	G-134		100			100				100	100					100
Jun. 2005	Acari	G-147			200	100				300	2400		300	600	200	300	
Jun. 2005	Acari	G-207			3400		100	900			300		100	200	100	2900	
Jun. 2005	Araneae	G-229			200			100			100		100			300	
Jun. 2005	Acari	G-244						100			1200			100			
Jun. 2005	Acari	G-262			1400						100						
Jun. 2005	Pseudosc.	G-263			100			100		100	100		100			100	
Jun. 2005	Acari	G-269															300
Jun. 2005	Pseudosc.	G-001		100													
Jun. 2005	Acari	G-100				100											
Jun. 2005	Acari	G-188														100	
Sep. 2005	Acari	G-002		800	800			300	100	200		200	200	100	300	1100	1500

Date	Order	Morpho	*Coccoloba*			*Conocarpus*			*Ficus*			*Pisonia*			*Tabebuia*		
		Type	L	O	H	L	O	H	L	O	H	L	O	H	L	O	H
Sep. 2005	Acari	G-003		400	700		100	500			1400			300		400	300
Sep. 2005	Acari	G-007		400	600		600	500	200	100	600		800	2600		300	300
Sep. 2005	Acari	G-031		900	400	100	600	1100	200	1200	1000	500	1100	1000	300		700
Sep. 2005	Acari	G-037		4400	2000	100	2100	1600	300	600	1300		1000	1600	200	900	800
Sep. 2005	Acari	G-078		1100	3100		400	1300	100	400	3900	100	600	2200	200	800	3900
Sep. 2005	Acari	G-105	100	1300	200		700		1400	700	500	800	1700	100	700	5500	300
Sep. 2005	Acari	G-147		100			200	300	100	300	700		400	1300	100	400	600
Sep. 2005	Acari	G-188		200			200		400	100	300		100	1200	700	300	300
Sep. 2005	Acari	G-207		500	400		200	400	100	300	200	100	100	500	800	900	2100
Sep. 2005	Acari	G-244		100	100		100	1300	100	200	500		300	2800		500	900
Sep. 2005	Pseudosc.	G-001								100	100		200				
Sep. 2005	Acari	G-004			200			600		100	200		100	700			
Sep. 2005	Isopoda	G-087			100	200		600		300							
Sep. 2005	Araneae	G-134	100			100	100					100					
Sep. 2005	Araneae	G-229					100			200	100				100	100	
Sep. 2005	Acari	G-262															1100
Sep. 2005	Pseudosc.	G-263		300	100								100	100			
Sep. 2005	Acari	G-288			200			200									100
Sep. 2005	Isopoda	G-010															
Sep. 2005	Psocoptera	G-102								100			100				
Sep. 2005	Diptera	G-123	100									100					

Date	Order	Morpho	*Coccoloba*			*Conocarpus*			*Ficus*			*Pisonia*			*Tabebuia*		
		Type	L	O	H	L	O	H	L	O	H	L	O	H	L	O	H
Sep. 2005	Diptera	G-182	100			100			100								
Sep. 2005	Pseudosc.	G-220				100											
Sep. 2005	Blattodea	G-228							100								
Sep. 2005	Acari	G-269	100														100
Sep. 2005	Diptera	G-157														100	
Sep. 2005	Diptera	G-191	100														
Sep. 2005	Diptera	G-209	100														

Author details

María F. Barberena-Arias[1*] and Elvira Cuevas[2]

1 Universidad Ana G. Méndez, Gurabo, Puerto Rico

2 University of Puerto Rico, San Juan, Puerto Rico

*Address all correspondence to: mbarberena1@uagm.edu

References

[1] Eviner VT, Chapin III FS. FUNCTIONAL MATRIX: A Conceptual Framework for Predicting Multiple Plant Effects on Ecosystem Processes. Annu. Rev. Ecol. Evol. Syst. 2003;34: 455–485. DOI: 10.1146/annurev.ecolsys.34.011802.132342

[2] Van Miegrot H, Hysell MT, Johnson AD. Soil Microclimate and Chemistry of Spruce–Fir Tree Islands in Northern Utah. Soil Science Society of America. 2000;64:1515-1525. DOI: 10.2136/sssaj2000.6441515x

[3] Haworth K, McPherson GR. Effects of *Quercus emoryi* trees on precipitation distribution and microclimate in a semi-arid savanna. Journal of Arid Environments. 1995;31: 153–170. DOI: 10.1006/jare.1995.0057

[4] Belsky AJ, Amundson RG, Duxbury JM, Riha SJ, Ali AR, Mwonga SM. The Effects of Trees on Their Physical, Chemical and Biological Environments in a Semi-Arid Savanna in Kenya. Journal of Applied Ecology. 1989:26(3):1005-1024. DOI: 10.2307/2403708

[5] Saxena M. Microclimate modification: calculating the effect of trees on air temperature. In: Proceedings of the solar conference; 21-25 April 2001; Washington DC

[6] Lugo AE, McCormick JF. Influence of environmental stressors upon energy flow in a natural terrestrial ecosystem. In: Barrett GW, Rosenberg R, editors. Stress effects on natural ecosystems: Wiley; 1981. P.79-102. DOI: 10.1017/S0376892900015411

[7] Hogg EW, Leiffers VJ. The impact of *Calamagrostis canadensis* on soil thermal regimes after logging in northern Alberta. Canadian Journal of Forest Research. 1991;21:387–394. DOI: 10.1139/x91-048.

[8] Bullock SH, Mooney HA, Medina E. Seasonally dry tropical forests. 1st ed. Cambridge University Press; 1995. 439 p. DOI: 10.1017/CBO9780511753398.001.

[9] Cuevas E, Lugo AE. Dynamics of organic matter and nutrient return from litterfall in stands of ten tropical tree plantation species. Forest Ecology and Management. 1998;112(3):263-279. DOI: 10.1016/S0378-1127(98)00410-1

[10] Steltzer H, Bowman WD. Original Articles: Differential Influence of Plant Species on Soil Nitrogen Transformations Within Moist Meadow Alpine Tundra. Ecosystems. 1998;1:464–474. DOI: 10.1007/s100219900042

[11] Swift MJ, Heal OW, Anderson JM. Decomposition in Terrestrial Eco-systems, volumen 5. 1st ed. University of California Press; 1979. 384 p.

[12] Hansen RA. Diversity in the decomposing landscape. In: Coleman D, Hendrix P, editors. Invertebrates as webmasters in ecosystems. CABI; 2000. P. 203-219. DOI: 10.1079/9780851993942.0000

[13] Ettema CH, Wardle DA. Spatial soil ecology. Trends in Ecology & Evolution. 2002;17(4):177-183. DOI: 10.1016/S0169-5347(02)02496-5

[14] Wardle DA. Communities and Ecosystems: Linking the Aboveground and Belowground. 1st edition. Princeton University Press; 2002; 387 p.

[15] Wardle DA, Lavelle P. Linkages between soil biota, plant litter quality and decomposition. In: Cadish G, Giller KE, editors. Driven by nature: plant litter quality and decomposition. CAB International; 1997. P. 107-123.

[16] 1999 Chen B, Wise DH. Bottom-up limitation of predaceous arthropods in a

detritus-based terrestrial food web. Ecology. 1999;80:761-772. DOI: 10.2307/177015.

[17] González G, Seastedt TR, Donato Z. Earthworms, arthropods and plant litter decomposition in aspen (*Populus tremuloides*) and lodgepole pine (*Pinus contorta*) forests in Colorado, USA. Pedobiologia. 2003;47:863-869. DOI: 10.1078/0031-4056-00272.

[18] González G, Zou X. Plant and litter influences on earthworm abundance and community structure in a tropical wet forest. Biotropica. 1999:31: 486-493. DOI: 10.1111/j.1744-7429.1999.tb00391.x

[19] Dilly O, Irmler U. Succession in the food web during decomposition of leaf litter in a black alder (*Alnus glutinosa* (Gaertn.) L.) forest. Pedobiologia. 1998: 42:109-123.

[20] Irmler U. Changes in the fauna and its contribution to mass loss and N release during leaf litter decomposition in two deciduous forests. Pedobiologia. 2000;44:105-118.

[21] Hasegawa M. Changes in Collembola and Cryptostigmata communities during the decomposition of pine needles. Pedobiologia. 1997;41: 225-241.

[22] Berg MP, Kniese JP, Bedaux JJM, Verhoef HA. Dynamics and stratification of functional groups of micro- and meso- arthropods in the organic layer of a Scots pine forest. Biology and fertility of soils 1998;26: 268-284. DOI: 10.1007/s003740050378

[23] Rivera-Rivera MJ, Acosta-Martínez V, Cuevas E. Tree Species and Precipitation Effect on the Soil Microbial Community Structure and Enzyme Activities in a Tropical Dry Forest Reserve. In: . Extremophilic Microbes and Metabolites - Diversity, Bioprespecting and Biotechnological Applications. 2020. DOI: 10.5772/intechopen.82579

[24] Barberena-Arias MF, Cuevas E. Physicochemical Foliar Traits Predict Assemblages of Litter/ Humus Detritivore Arthropods. In: Tropical Forests - New Edition. Sudarshana P, Nageswara-Rao M, Soneji JR, editors. IntechOpen. 2018. DOI: 10.5772/intechopen.75076.

[25] Hooper DU, Chapin III FS, Ewel JJ, Hector A, InchaustiP, Lavorel S, Lawton JH, Lodge DM, Loreau M, Naeem S, Schmid B, Setälä H, Symstad AJ, Vandermeer J, Wardle DA. Effects of biodiversity on ecosystem functioning: a consensus of current knowledge. Ecological Monographs. 2005:75(1):3-35. DOI: 10.1890/04-0922

[26] Ebeling A, Rzanny M, Lange M, Eisenhauer N, Hertzog LR, Meyer ST, Weisser WW. Plant diversity induces shifts in the functional structure and diversity across trophic levels. Oikos. 2018;127(2):208-219. DOI: 10.1111/oik.04210

[27] Prather RM, Castillioni K, Welti EAR, Kaspari M, Souza L. Abiotic factors and plant biomass, not plant diversity, strongly shape grassland arthropods under drought conditions. Ecology. 2020;101(6):e03033. 10.1002/ecy.3033

[28] Welti E, Helzer C, Joern A. Impacts of plant diversity on arthropod communities and plant–herbivore network architecture. Ecosphere. 2017;8 (10):e01983. DOI: 10.1002/ecs2.1983

[29] De Deyn GB, Raaijmakers CE, Van Ruijven J, Berendse F, Van Der Putten WH. Plant species identity and diversity effects on different trophic levels of nematodes in the soil food web. Oikos. 2004;106(3):576-586. DOI: 10.1111/j.0030-1299.2004.13265.x

[30] Díaz S, Hodgson JG, Thompson K, Cabido M, Cornelissen JHC, Jalili A,

Montserrat-Martí G, Grime JP, Zarrinkamar F, Asri Y, Band SR, Basconcelo S, Castro-Díez P, Funes G, Hamzehee B, Khoshnevi M, Pérez-Harguindeguy N, Pérez-Rontomé MC, Shirvany A, Vendramini F, Yazdani S, Abbas-Azimi R, Bogaard A, Boustani S, Charles M, Dehghan M, de Torres-Espuny L, Falczuk V, Guerrero-Campo J, Hynd A, Jones G, Kowsary E, Kazemi-Saeed F, Maestro-Martínez M, Romo-Díez A, Shaw S, Siavash B, Villar-Salvador P, Zak MR. The plant traits that drive ecosystems: evidence from three continents. Journal of vegetation science, 2004;15: 295-304. DOI: 10.1111/j.1654-1103.2004.tb02266.x

[31] Rivera-Rivera MJ, Cuevas E. First Insights into the Resilience of the Soil Microbiome of a Tropical Dry Forest in Puerto Rico. In: Blumenberg M, Shaaban M, Elgaml A. Microorganisms. IntechOpen; 2020. DOI: 10.5772/intechopen.90395.

[32] Murphy PG, Lugo AE. Dry forests of the subtropics: Guánica forest in context. Acta científica. 1990;4:15-24.

[33] Lugo AE, Ramos O, Molina S, Scatena FN, Vélez-Rodríguez LL. A fifty-three year record of land use change in the Guánica forest Biosphere Reserve and its vicinity. IITF, USDA Forest Service, Rio Piedras, PR 13. 1996. p + maps + photos.

[34] Dunphy BK, Murphy PG, Lugo AE. The tendency for trees to be multiple-stemmed in tropical and subtropical dry forests: Studies of Guanica forest, Puerto Rico. Tropical Ecology. 2000;41 (2): 161-167.

[35] Medina E, Cuevas E. Propiedades fotosintéticas y eficiencia de uso de agua de plantas leñosas del bosque deciduo de Guánica: consideraciones generales y resultados preliminares. Acta Científica. 1990;4:25-36.

[36] Lugo AE, González-Liboy JA, Cintrón B, Dugger K. Structure, productivity and transpiration of a subtropical dry forest in Puerto Rico. Biotropica. 1978;10:278-291. DOI: 10.2307/2387680

[37] Cintrón B, Lugo AE. Litter fall in a subtropical dry forest: Guánica, Puerto Rico. Acta científica. 1990;4:37-50.

[38] Farnsworth B. A guide to trails of Guánica. State forest and biosphere reserve. Departamento de recursos naturales de Puerto Rico. 1991.

[39] Medina E, Cuevas E, Molina S, Lugo AE, Ramos O. Structural variability and species diversity of a dwarf Caribbean dry forest. Caribbean Journal of Science. 2010;46(2-3): 203-215.

[40] NRCS National Water and Climate Center [Internet]. 2008. Available from http://www.wcc.nrcs.usda.gov/scan/site.pl?sitenum=2067&state=pr

[41] RAWS Remote Automated Weather Stations - Western Regional Climate [Internet]. 2008. Available from http://www.wrcc.dri.edu/wraws/prF.html.

[42] Barberena-Arias MF, González G, Cuevas E. Quantifying Variation of Soil Arthropods Using Different Sampling Protocols: Is Diversity Affected? In: Sudarshana P, Nageswara-Rao M, Soneji JR, editors. Tropical Forests. IntechOpen; 2012. DOI: 10.5772/30386.

[43] McCune B, Grace JB. Analysis of Ecological Communities. 1st ed. Gleneden Beach: MjM Software Design; 2002. 300 p. ISBN: 0972129006

[44] Legendre P, Legendre L. Numerical Ecology—Developments in Environmental Modeling 20. 1st ed. New York: Elsevier; 1998. ISBN: 0444892508

[45] Peck J. Multivariate Analysis for Community Ecologists. 1st ed. Gleneden

Beach: MjM Software Design; 2010. 162 p. ISBN: 9780972129022

[46] Murphy PG, Lugo AE. Ecology of tropical dry forest. Annual Review of Ecology and Systematics. 1986;17:67-88. DOI: 10.1146/annurev.es.17.110186.000435

[47] Pressland AJ. Rainfall partitioning by an arid woodland (Acacia aneura F. Muell.) in south-western Queensland. Australian Journal of Botany. 1973;21: 235-245. DOI: 10.1071/BT9730235

[48] Clauser C, Huenges E. Thermal conductivity of rocks and minerals. In: Ahrens TJ, editor. Rock Physics and Phase Relations, A Handbook of Physical Constants, AGU Reference Shelf 3. 1995. P. 105-126. DOI: 10.1029/RF003p0105

[49] Barberena-Arias MF. Single tree species effects on temperature, nutrients and arthropod diversity in litter and humus in the Guánica dry forest [thesis]. San Juan PR, Univesity of Puerto Rico. 2008.

[50] Ecosystem feedbacks on climate at the landscape scale. Philosophical Transactions of the Royal Society London B. 1998;353:5-18.

[51] Malmgrem BA, Winter A. Climate zonation in Puerto Rico based on principal components analysis and an artificial neural network. Journal of climate. 1999;12:977-985. DOI: 10.1175/1520-0442(1999)012<0977:CZIPRB>2.0.CO;2

[52] Rosenberg NJ. Microclimate: the biological environment. Wiley Interscience; 1974. 528 p.

[53] Vandenbeldt RJ, Williams JH. The effect of soil surface temperature on the growth of millet in relation to the effect of *Faidherbia albida* trees. Agricultural and forest meteorology. 1992;60(1-2): 93-100. DOI: 10.1016/0168-1923(92)90076-G

[54] Axelrod F. A systematic vademecum to the vascular plants of Puerto Rico, Herbarium Manager UPRRP. 2007.

[55] Loigier AH. La flora de la Espanola, Universidad Central del Este, San Pedro de Macoris, R.D. Vols. I – VIII, Serie Cientifica 29. 1996

[56] Prinzing AP, Lentzsch P, Voigt F, Woas S. Habitat stratification stratifies a local population: Ecomorphological evidence from a bisexual, mobile invertebrate (Carabodes labyrinthicus; Acari). Annales Zoologici Fennici. 2004; 41(2):399-412. ISSN: 0003-455X

[57] Kardol P, Cornips NJ, van Kempen MM, Bakú-Schotman JMT, van der Putten WH. Microbe-mediated plant-soil feedback causes historical contingency effects in plant community assembly. Ecological Monographs. 2007; 77(2):147-162. DOI: 10.1890/06-0502

[58] Bezemer TM, Lawson CS, Hedlund K, Edwards AR, Brooks AJ, Igual JM, Mortimer SR, van der Putten WH. Plant species and functional group effects on abiotic and microbial soil properties and plant-soil feedback responses in two grasslands. Journal of Ecology. 2006;94:893-904. DOI: 10.1111/j.1365-2745.2006.01158.x

[59] Herrera F. Dinámica especial y temporal de la fauna del suelo en un mosaico de vegetación de la cordillera de la costa [tesis]. Caracas: IVIC Venezuela. 1997.

Chapter 8

Mixed Diets Enhance Edible Grasshopper, *Ruspolia differens* (Orthoptera: Tettigoniidae) Performance during Mass Rearing

Geoffrey Maxwell Malinga, Robert Opoke and Karlmax Rutaro

Abstract

Mixing of diets is a notable dietary practice that is believed to improve performance-related characteristics such as growth, survival rate and egg-laying potential among insect herbivores. However, currently there is limited information regarding the performance of edible insects either on artificial and natural diets or their mixtures. This chapter reviewed recent literature on performance of a seasonally harvested and a widely consumed edible grasshopper, *Ruspolia differens* (Orthoptera: Tettigoniidae) reared on various artificial and natural diets. Our aim is to highlight diets and diet mixtures that results in the highest *R. differens* production. The results of the review show that *R. differens* performs better on mixed diets than on single or less diversified diets. In all reviewed studies, edible grasshoppers fed mixed diets either of natural plants or artificial diets achieved highest final weights, highest survival, highest fecundity and fastest development times than less diversified diets. The information is useful in designing technologies for large-scale rearing program for this species.

Keywords: insect farming, developmental time, captive rearing, African edible bush cricket

1. Introduction

Diet mixing, i.e., feeding on more than one plant species or food resource is a well-known feeding habit among polyphagous insect herbivores and is believed to be associated with increased growth, shorter development time, survival rate and fecundity [1, 2]. This is because mixed diets could allow nutrient complementation and reduce the amounts of toxins in solitary foods (reviewed in [3, 4]). Several polyphagous orthopteran species have been shown to enhance their performance by diet mixing and exhibit higher fitness and survival rates, and faster growth on mixed diets than on single resource diets [1, 2, 5–7]. However, despite this widespread evidence of diet mixing in insect herbivores, in contrast, very little has been done to review the effect of diet mixtures on the performance of edible insects, which hamper the development of a suitable diet or diet mixtures for sustaining mass-rearing.

IntechOpen

The edible grasshopper, *Ruspolia differens* (Orthoptera: Tettigoniidae) is a seasonally harvested edible insect species. This edible grasshopper species is largely found in tropical Africa particularly in Kenya, Angola, Uganda, South Africa, Ghana, Central African Republic, Democratic Republic of Congo, Zambia, Rwanda, Burundi, Mauritius, Madagascar, Ivory Coast, Tanzania, where it is eaten as a delicious food [8–15]. It is nutritionally rich and healthy. *Ruspolia differens* have been found to have high levels of protein (34-73%), fat content (33-48%), carbohydrates (3-6%), polyunsaturated fatty acids (89% of lipids), chitin (10-15%) and several minerals including phosphorous (141-673 mg/100 g), potassium (446-673 mg/100 g), calcium (34.9-128 mg/100 g), Fe (17 mg/100 g) and zinc (12-17 mg/100 g). This katydid grasshopper is predominantly nocturnal, showing distinctive colour polymorphism and sex dimorphism [8]. There are six colourmorphs comprising of green, purple-stripped green, purple-suffused green, brown, purple-stripped brown and purple-suffused-brown [16] although the green and brown colourmorphs tend to dominate. *Ruspolia differens* is characterised by a unique swarming phase that exist during rainy seasons and a non-swarming phase (also called local population) whose developmental stages are found in the field all over the year [15, 17]. In East Africa, swarming occurs twice annually in April–May and November–December [8], and the source populations of the swarming individuals are believed to originate locally [17]. Male adults are smaller than females, and are characterised by longer antenna and possess a pair of tongue-like metathoracic flaps and dorsoventrally bi-lobed cerci, while, females are identified by the presence of a long slender ovipositor [10, 15, 16, 18–21]. The life cycle is variable but roughly 147 days and there are six nymphal stages for male and seven for females [12]. Eggs are oviposited in the leaf sheaths of grasses, shoots of Panicum sp., maize seedlings and *Pennisetum sp*, and cotton wool and folded plastic cloth [22–24]. Hatching happens after 11-25 days and is determined by the moisture level [22, 24, 25]. *Ruspolia differens* is a facultatively oligophagous grass-feeder and can feed on a range of grass and sedge species [17, 26, 27] and various artificial diets [12, 25, 28]. However, the species is highly selective preferring inflorescences over leaves [27] but they are flexible if a limited range of diets are available [27, 28]. Despite the large number of recent research works on *R. differens*, however, there is dearth of information regarding the performance of edible insects either on artificial and natural diets or their mixtures. This chapter systematically reviewed all recent peer-reviewed literature on the development and reproductive performance of edible grasshopper, *Ruspolia differens* (Orthoptera: Tettigoniidae) reared on various artificial and natural diets over the last 20 years. Our aim is to highlight diets and diet mixtures that results in the highest *R. differens* production. The information is useful in designing large-scale rearing program for this species.

2. Materials and methods

A comprehensive literature search was conducted within the electronic databases such as Scopus, Web of Science and Google Scholar to identify peer-reviewed journal articles focussing on performance of *R. differens* on mixed or diversifying diets. In the search, following key words were used to demarcate the species in the search: *Ruspolia differens*, Long-horned grasshopper, Edible bush cricket, Swarming conehead, Mixed diets and Performance. The inclusion criteria included only peer-reviewed papers published within the last 20 years (from 2000 to 2020, the period of active research on *R. differens*). Because of the aggravated risk of bias as a result of translation of information from other languages to English, our search was confined to English language only. Dissertation, conference proceedings and papers published in other languages were excluded from the review. The title, abstracts and full texts of the articles were screened against the selection criteria. The search yielded six articles out of which three were relevant and were included in the systematic review. The statistical assessments of effects of dietary treatment on development time (days) from the first instar to adult and fresh weight at adult stage and female fecundity were done using Linear Mixed model with type III sum of squares (sex and diet-sex interaction included as fixed factors and block as a random variable) followed by Bonferroni post hoc pairwise comparisons. A logistic regression model was used to determine whether diet treatment (predictor variable) predicts the nymphal survival to adulthood (outcome variable) with block included as a random variable. All statistical analyses were conducted using IBM SPSS Statistics software version 26. For artificial diets study, Linear contrasts within the analysis of variance (ANOVA) were employed to examine the existence of a linear trend in the insect performance variables along the food resource diversity gradient.

3. Results

The reviewed articles were based on recent studies conducted over the last 20 years (2000 to 2020) in Uganda and Kenya. Most of these articles were published in 2018. The studies focused on identification of articles that focused on performance of edible grasshopper, *R. differens* on either natural or artificial diets.

3.1 Performance of *R. differens* on mixed artificial diets

The development of a successful and sustainable mass rearing strategy requires knowledge of potentially suitable diet mixtures which can best support the insect's growth, development and survival. A study by Malinga et al. [25] in Uganda, examined how diet treatments representing a gradient of increasing food resource diversity, ranging from one food (single-food treatment) to mixtures of two, three, five, six and eight foods (**Table 1**), affect the developmental and reproductive performance of the edible grasshopper, *R. differens*. In all reviewed studies, approximately equal amounts of foods were administered ad-libitum to each individual in the diet mixture till moulting to adult, and water was provided either by introduction of moistened tissue paper or by spraying droplets of water with a portable hand sprayer.

More diversified diets resulted in shorter development time rather than in the single diet or less diversified diets and there was a significant decreasing linear trend in the development time along the food resource diversity gradient, indicating that as food resource diversity increases from single to more diversified diet mixtures, development time decreases proportionally. The improved performance are likely due to the utilisation of a balance of high-quality nutrients in mixed diets [2]. Benefits of diet mixing on insect performance (e.g., developmental time) has previously been documented also on several generalist Acrididae grasshoppers [1–3, 5, 29]. These studies have found generalist grasshopper to perform better on high quality foods which in turn determines their growth and developmental rate [30]. Increasing dietary diversification resulted in greater adult weight and there was a significant increasing linear trend in the adult fresh weight along the food resource diversity gradient. Furthermore, the experimental results of this work indicated that dietary mixing greatly improved female fecundity (total number of eggs laid per female, on average roughly 90-140 eggs). There was also a significant increasing linear trend in the female fecundity along the food resource gradient. Improved

Treatment code	Treatment name	Composition
One	Single food treatment	Rice seed head
Two	Two food mixtures	Rice seed head
		Finger millet seed head
Three	Three food mixtures	Rice seed head
		Finger millet seed head
		Wheat bran
Five	Five food mixtures	Rice seed head
		Finger millet seed head Wheat bran
		Chicken superfeed egg booster (from super Feeds Farm Products Ltd., Kampala, Uganda)
		Sorghum seed head
Six	Six food mixtures	Rice seed head
		Finger millet seed head Wheat bran Chicken superfeed egg booster Sorghum seed head Germinated finger millet
Eight	Eight food mixtures	Rice seed head
		Finger millet seed head
		Wheat bran
		Chicken superfeed egg booster Sorghum seed head Germinated finger millet
		Simsim cake
		Dog biscuit pellet

Table 1.
Compositions of diets used in the performance experiments.

fecundity in mixed diets has also been reported for the generalist grasshoppers *Parapodisma subastris* (Huang) [1] and *Chorthippus parallelus* (Zetterstedt) [2] (Orthoptera: Acrididae).

Finally, more diversified diets resulted in greatly improved survival rates (approximately 31-51%) of *R. differens* to adulthood than when reared on single resource diets (11%). Even with a minimal diet diversification, nymphal survival to adult was greatly improved. The odds of survival (the probability of surviving relative to dying) was four to eight times higher for individuals fed mixtures of artificial foods than in those fed one food only. Numerous studies have shown that enhanced survival and development in the majority of herbivorous insects correspond to diet quality and quantity on which the insects are fed [30–32]. For instance, Oonincx et al. [32] established that larvae of the black soldier fly, *Hermetia illucens* (L.) (Diptera: Stratiomyidae) reared on vegetable by-products diets high in protein achieved a relatively shorter development time (21 days) that their counterparts reared on low-protein diets (37 days). Thus, the remarkably lower survival observed on single than in mixed diets could reflect the low protein and high starch content in solitary foods, as also found previously with mealworms [29]. The results of our review demonstrate how a cost-effective optimal mass production of *R. differens* could be achieved even with minimal dietary mixtures.

3.2 *Ruspolia differens* performance on diet mixtures of natural food plants

In a study evaluating the suitability of inflorescences for rearing *R. differens* [26], individual *R. differens* were reared from first instar to adults on diet treatments representing a gradient of five diversifying dietary mixtures of host plant species ranging from one to mixtures of two, three, five and seven species (**Table 2**). The gradient represented the hierarchy of use, based on a survey of potential food plants of *R. differens* used in the field in Uganda [17], whereby *Brachiaria ruziziensis* was the most used and *B. ruziziensis* and *P. maximum* were the two most used host plants in the field, respectively.

Treatment code	Treatment name	Composition
One	One host plant	*Brachiaria ruziziensis*
Two	Two host plants mixture	*B. ruziziensis*, *Panicum maximum*
Three	Three host plants mixture	*B. ruziziensis*, *P. maximum*, *Hyparrhenia rufa*
Five	Five host plants mixture	*B. ruziziensis*, *P. maximum*, *H. rufa*, *Chloris gayana*, *Cynodon dactylon*
Seven	Seven host plants mixture	*B. ruziziensis*, *P. maximum*, *H. rufa*, *Ch. gayana*, *Cy. dactylon*, *Sporobolus pyramidalis*, *Pennisetum purpureum*

Table 2.
Host plant combinations used in mixtures.

The rearing containers were placed in blocks, each containing one replicate of each host plant with one *R. differens* individual per container. In each container, approximately equal quantities of foods were randomly placed. The results showed that dietary mixture of grass species inflorescences is beneficial for the survival and for shortening the development time of the edible *R. differens* but not for achieving a higher adult emergence weight. For example, the nymphal developmental time was significantly shorter (on average 16 days shorter) when *R. differens* individuals were reared on more diverse diets than when reared on single diets (>90 days for diets with 1-2 plant species; <80 days for diets with 5-7 plants). Similarly, the survival to adult stage was higher in the most diversified diet mixture than in the least diversified diets, ranging from 12.5-15% in the three least diversified diet treatments to 40-65% in the two most diversified diets with an overall average nymphal survival of 30%. In another study conducted by Ssepuuya et al. [33], highest survival of *R. differens* was found after eight weeks on mixed diets comprising of green stems and leaves of three plants *C. dactylon*, *P. maximum* and *E. africana*) than in single plant diets. Consistent with our reviewed findings, previous works on generalist grasshoppers, *P. subastris* (Huang) [1] and *C. parallelus* (Zetterstedt) [2] also recorded highest survival in the 6-and 8-food plant (grass species) mixtures, respectively, than in food treatments with only a single plant species. The better performance on diet mixtures may be driven by a better gain of nutrients amounts in mixed than in single host plants [2].

4. Conclusion and future prospects

To conclude, the results of this review collectively provide evidence that *R. differens* performs better on mixed diets than on single or less diversified diets.

These results emphasise the importance of diet mixing in the optimisation of edible grasshopper, *R. differens*, production during mass-rearing programs. However, to upscale from current laboratory rearing to full-scale mass-rearing programmes suitable for local farmers or entrepreneurs, there are still several knowledge-gaps and steps remaining, e.g., what are the most cost-effective feeds in mass-rearing, and could side-streams from East-African food-industry or agricultural farms be included in feeds (reducing the costs and enhancing environmental sustainability of the rearing)? Do the used feeds in rearing modify the taste of *R. differens*? Economics of rearing: what feed and feeding regimes are economically viable?. Furthermore, research is needed to determine the combined effects of diet mixtures of natural and artificial foods on the performance of *R. differens*.

Author details

Geoffrey Maxwell Malinga[1*], Robert Opoke[2] and Karlmax Rutaro[3]

1 Department of Biology, Gulu University, Gulu, Uganda

2 Department of Biology, Muni University, Arua, Uganda

3 Department of Biochemistry and Sports Science, Makerere University, Kampala, Uganda

*Address all correspondence to: malingageoffrey@yahoo.com

References

[1] Miura, K., & Ohsaki, N. (2004). Diet mixing and its effect on polyphagous grasshopper nymphs. Ecological Research, 19(3), 269-274.

[2] Unsicker, S. B., Oswald, A., Köhler, G., and Weisser, W. W. (2008). Complementarity effects through dietary mixing enhance the performance of a generalist insect herbivore. Oecologia, 156(2), 313-324.

[3] Bernays, E. A., & Bright, K. L. (1993). Mechanisms of dietary mixing in grasshoppers: a review. Comparative Biochemistry and Physiology Part A: Physiology, 104(1), 125-131.

[4] Hägele BF, Rowell-Rahier, M. (1999) Dietary mixing in three generalist herbivores: nutrient complementation or toxin dilution? Oecologia 119:521-533.

[5] MacFarlane JH, Thorsteinson AJ (1980) Development and survival of the two stripped grasshoppers *Melanoplus bivittatus* (Say) (Orthoptera: Acrididae) on various single and multiple diets. Acrida 9:63-76.

[6] Singer, M.S. & Bernays, E.A. (2003) Understanding omnivory through food-mixing behaviour. Ecology 84, 2532-2537.

[7] Swan, C. M. and Palmer, M. A. (2006). Composition of speciose leaf litter alters stream detritivore growth, feeding activity and leaf breakdown. Oecologia, 147(3):469-478.

[8] Bailey, W. J. and McCrae, A. W. R. (1978). The general biology and phenology of swarming in the East African tettigoniid *Ruspolia differens* (Serville) (Orthoptera). Journal of Natural History, 12(3): 259-288.

[9] Okia, C. A., Odongo, W., Nzabamwita, P., Ndimubandi, J., Nalika, N., & Nyeko, P. (2017). Local knowledge and practices on use and management of edible insects in Lake Victoria basin, East Africa. Journal of Insects as Food and Feed, 3(2): 83-93.

[10] Mmari, M. W., Kinyuru, J. N., Laswai, H. S., & Okoth, J. K. (2017). Traditions, beliefs and indigenous technologies in connection with the edible longhorn grasshopper *Ruspolia differens* (Serville 1838) in Tanzania. Journal of Ethnobiology and Ethnomedicine, 13(1): 1-11.

[11] Massa B (2015a) Taxonomy and distribution of some katydids (Orthoptera: Tettigoniidae) from tropical Africa. ZooKeys 524:17-44.

[12] Brits, JH and Thornton, C. H. (1981). On the biology of *Ruspolia differens* (Serville)(Orthoptera: Tettigoniidae) in South Africa. Phytophylactica, 13(4):169-174.

[13] Matojo, N. D., & Hosea, K. M. (2013). Phylogenetic relationship of the longhorn grasshopper *Ruspolia differens* Serville (Orthoptera: Tettigoniidae) from Northwest Tanzania based on 18S ribosomal nuclear sequences. Journal of Insects, Volume 2013 |Article ID 504285 | https://doi.org/10.1155/2013/504285.

[14] Massa B (2015b) Taxonomy and distribution of some katydids (Orthoptera: Tettigoniidae) from tropical Africa. ZooKeys 524:17-44.

[15] Matojo, N. D. (2017). A review work on how to differentiate the longhorn grasshoppers Ruspolia differens and Ruspolia nitidula (Orthoptera: Tettigoniidae). Journal of Applied Life Sciences International, 1-4.

[16] Matojo ND (2020). A Comprehensive Key for Identification of the "Swarming Conehead" Ruspolia Differens Serville, 1838 (Orthoptera: Tettigoniidae) Occurring in the

Afro-Tropical Region. International Journal of Zoology and Animal Biology, 3(1): 000200. DOI: 10.23880/izab-16000200.

[17] Opoke R, Nyeko P, Malinga GM, Rutaro K, Roininen H, Valtonen A (2019) Host plants of the non-swarming edible bush cricket Ruspolia differens. Ecology and Evolution 9: 3899-3908.

[18] Matojo DN, Yarro GJ (2010) Variability in polymorphism and sex ratio of the conehead Ruspolia differens Serville (Orthoptera: Conocephalidae) in north-west Tanzania. Int J Integr Biol 9:131-136.

[19] Matojo DN, Njau MA (2010) Plasticity and biosystematics of swarming of the conehead *Ruspolia differens* Serville (Orthoptera: concephalidae). International Journal of Integrated Biology 9(97):231-243.

[20] Matojo, N. D., & Yarro, J. G. (2013). Anatomic Morphometrics of the "Senene" Tettigoniid *Ruspolia differens* Serville (Orthoptera: Conocephalidae) from North-West Tanzania. International Scholarly Research Notices, Volume 2013 |Article ID 176342 | https://doi.org/10.1155/2013/176342.

[21] Ssepuuya, G., Nakimbugwe, D., De Winne, A., Smets, R., Claes, J., & Van Der Borght, M. (2020). Effect of heat processing on the nutrient composition, colour, and volatile odour compounds of the long-horned grasshopper *Ruspolia differens* Serville. Food Research International, 129:108831. https://doi.org/10.1016/j.foodres.2019.108831.

[22] Egonyu, J. P., Miti, M. M., Tanga, C. M., Alfonce, L., & Subramanian, S. S. (2020). Cannibalism, oviposition and egg development in the edible long-horned grasshopper, *Ruspolia differens* (Orthoptera Tettigoniidae) under laboratory conditions. Journal of Insects as Food and Feed: 7 (1):89-97.

[23] Malinga, G. M., Lehtovaara, V. J., Valtonen, A., Nyeko, P., & Roininen, H. (2019). Developing mass egg-laying medium for the edible Ruspolia differens (Orthoptera: Tettigonidae). Journal of Economic Entomology, 112(5): 2157-2160.

[24] Hartley JC (1967) Laboratory culture of a Tettigoniid, *Homorocoryphus nitidulus vicinus* (Wlk.) (Orthoptera). Bulletin of Entomological Research 57: 203-205.

[25] Malinga GM, Valtonen A, Lehtovaara VJ, Rutaro K, Opoke R, Nyeko P, Roininen H (2018b) Mixed artificial diets enhance the developmental and reproductive performance of the edible grasshopper, *Ruspolia differens* (Orthoptera: Tettigoniidae). Applied Entomology and Zoology 53: 237-242.

[26] Malinga, G. M., Valtonen, A., Hiltunen, M., Lehtovaara, V. J., Nyeko, P., & Roininen, H. (2020). Performance of the African edible bush-cricket, *Ruspolia differens*, on single and mixed diets containing inflorescences of their host plant species. Entomologia Experimentalis et Applicata, 168(6-7):448-459.

[27] Valtonen, A., Malinga, G. M., Junes, P., Opoke, R., Lehtovaara, V. J., Nyeko, P., & Roininen, H. (2018). The edible katydid *Ruspolia differens* is a selective feeder on the inflorescences and leaves of grass species. Entomologia Experimentalis et Applicata: 166(7): 592-602.

[28] Malinga, G. M., Valtonen, A., Lehtovaara, V. J., Rutaro, K., Opoke, R., Nyeko, P., & Roininen, H. (2018a). Diet acceptance and preference of the edible grasshopper *Ruspolia differens* (Orthoptera: Tettigoniidae). Applied Entomology and Zoology, 53(2):229-236.

[29] Van Broekhoven, S., Oonincx, D. G., Van Huis, A., & Van Loon, J. J. (2015).

Growth performance and feed conversion efficiency of three edible mealworm species (Coleoptera: Tenebrionidae) on diets composed of organic by-products. Journal of Insect Physiology, 73: 1-10.

[30] Joern A, Behmer ST (1997) Importance of dietary nitrogen and carbohydrates to survival, growth and reproduction in adults of the grasshoppers *Ageneotettix deorum* (Orthoptera: Acrididae). Oecologia 112:201-208.

[31] Nguyen, T. T., Tomberlin, J. K., & Vanlaerhoven, S. (2013). Influence of resources on *H. illucens* (Diptera: Stratiomyidae) larval development. Journal of Medical Entomology, 50(4):898-906.

[32] Oonincx, D. G. A. B., Van Huis, A., & Van Loon, J. J. A. (2015). Nutrient utilisation by black soldier flies fed with chicken, pig, or cow manure. Journal of Insects as Food and Feed, 1(2):131-139.

[33] Ssepuuya, G., C. M. Tanga, I. Yekko, F. Sengendo, C. T. Ndagire, K. K. M. Fiaboe, J. Karungi, and Nakimbugwe, D. (2018). Suitability of egg hatching conditions and commonly available food plants for rearing the long-horned grasshopper *Ruspolia differens* Serville (Orthoptera:Tettigoniidae). Journal of Insects as Food and Feed, 4(4): 253-261.

Chapter 9

Crustacea: The Increasing Economic Importance of Crustaceans to Humans

Gregorius Nugroho Susanto

Abstract

Crustaceans (subphylum Crustacea) are members of the phylum Arthropods, including crabs, lobsters, crayfish, prawn, shrimp, krill, barnacles, woodlice and beach fleas. The most common types of crustaceans are shrimp and crab. This subphylum is distinguished from other arthropods, including myriapods, insects, and chelicerates, by the presence of two-parted (biramous) appendages, and the hatchling's nauplius shape. In addition, these arthropods are majorly aquatic, often found in fresh, marine, or brackish water bodies, however, some crabs, hermit crabs, woodlice and other members of the subphylum, are found in terrestrial environments. Also, most crustaceans are free-living while numerous are parasitic (for instance, *Rhizocephala*, tongue worms, fish lice) and sessile (barnacles). Mostly lived nocturnal. Crustaceans have a great economic importance to humans. The group is of great value directly or indirectly for his health and economic progress, such as aesthetic, commercial, gastronomic, biomedical, bioindicator, biomonitor, geological values, and miscellaneous uses, biodeterioration and poisons.

Keywords: crustaceans, crustacea, arthropods, economic, humans

1. Introduction

Crustaceans (Crustacea) are a large group of arthropods, 4th largest diversity among the animal groups and are usually considered as a subphylum. This group comprise approximately 50,000 to 75,000 species and include many familiar animals such as crabs, lobsters, prawn, barnacle, woodlice and beach fleas, as well as a host of lesser-known species. Unlike the terrestrial Hexapoda and Myriapoda and mainly terrestrial Chelicerata, the main radiation of the Crustacea has been aquatic, with the bulk of species living in marine habitats. There is also a substansial number of freshwater species, but only 2–3% of species live on land. Crustaceans are the dominant arthropods in the oceans, where they occupy benthic, pelagic, planktonic and intertidal niches and lead motile, sedentary, sessile or parasitic life styles. In inland waters they are represented by a more limited range of taxa, but nevertheless have succeeded in virtually all types of water bodies, including freshwater, temporary pools, and even hypersaline lake [1]. On other hand, on land the diversity is low, with representatives from only three orders of malacostracans and poorly studied cryptozoic fauna of microcrustaceans. The crustaceans show an enormous diversity of form, and a great range of size from a minute planktonic and larval forms of

IntechOpen

Stygotantulus stocki at 0.1 mm (0.004 in) to giant crab, the Japanese spider crab, a large benthic crabs with a leg span of up to 12.5 ft. (3.8 m) and a mass of 48 lb. (20 kg), and also lobsters which can weigh up to 60 kg. The largest terrestrial species is the robber crab, *Birgus latro*, which can reach 3 kg in weight [2, 3]. Decapod crustaceans are possibly the most widely known vertebrates, due to the greatly cherished edibility. In this regard, shrimps, crabs and lobsters are of the highest commercial value. A study by De Grave et al. [4] estimated a total of 14,756 species and 2,725 genera of extant decapods.

Meanwhile, there are about 3,047 species of shrimps and prawns worldwide, and these are grouped into Caridea (2,517 species), Sergestoidea (94 species), Penaeiodea (94 species), and Stenopodidae (60 species). The commercial shrimp species majorly belong to one of the 5 penaeidean families (Penaeidae, Aristeidae, Sicyoniidae, Sergestidae and Solenoseridae) or the 3 caridean families (Pandalidae, Palaemonidae, and Crangonidae) [4].

2. General description

Crustaceans have hard skin (shells) due to deposits calcium carbonate in the cuticle. All or some parts of the body contain the appendix biramous originals. Breathe with gills or the entire surface of the body. Antenna glands (green glands) or the maxilla gland are excretion tools. Except for certain types, crustaceans are generally dioecious, fertilization inside. Most incubating the egg. The initial type of crustacean larvae is basically the nauplius larvae that swim freely as plankton. However, a few peculiar biological features possessed by crustaceans are not documented. These include, anatomy and morphology (segmentation, extremities, cuticle), circulation, excretion, respiration, osmotic regulation, procreation and life history habits (precopula, breeding, hatchlings, moulting), abundance and distribution, as well as mode of life (in the benthos or as plankton).

A crustacean's body comprises of segments, arranged into three regions, the head or cephalon, the thorax, and the abdomen or pleon. In some organisms, the head and thorax are merged to form a cephalothorax, often secured an expansive carapace [5]. Furthermore, the hard exoskeleton in crustaceans offers protection, and ought to moulted for development to occur. Each somite is surrounded by a shell separated into dorsal tergum, ventral sternum and a lateral pleuron, and the different exoskeleton parts are often combined together.

In addition, every somite or body portion often carry a pair of jointed appendages, while the head bears two sets of antennae, as well as the maxillae mandibles and the thoracic regions carry the legs. These legs are often specialized as maxillipeds (*feeding legs*) and pereiopods (*walking legs*). Meanwhile, pleopods (*swimming legs*) are located on the abdomen, this in turn closes in a telson, bears the anus, and is often bordered by uropods, forming a tail fan [2, 6]. The subphylum's remarkable survival is partly due to the large number and assortment of appendages. These are ordinarily biramous (comprised of two parts), with the exception of the uniramous primary antennae. This biramous nature possibly originated in crustaceans or became lost by other arthropods because of evolution, however, the exact origin is unclear, as even trilobites have biramous limbs.

Crustaceans also possess an open circulatory system, and blood circulated to the haemocoel the heart, adjacent to the dorsum. The oxygen carrying pigment in Malacostraca is haemocyanin, while the counterpart in copepods, ostracods, barnacles and branchiopods, is haemoglobin. In addition, the alimentary canal comprises a straight tube, progressing into a spiral, and often contains a "gastric mill" similar to a gizzard, as well as two digestive glands for food absorption. Also, there are some kidney-like structures as well as a ganglia-shaped brain close to the antennae, and a group of major ganglia, beneath the intestine.

Numerous male decapods have primary (and in some cases, secondary) pairs of pleopods for transfer of sperm. Meanwhile, several terrestrial crustaceans (for instance, the Christmas Island red crab) mate often and migrate to the ocean to lay eggs, while others, including woodlice, lay eggs in soggy earth. Female decapods mostly carry eggs and give birth to free-swimming hatchlings.

3. Classification

Based on body size Crustaceans are grouped into the smaller forms called Entomostraca and the larger members called Malacostraca. Entomostraca is not considered a valid taxonomic division because the members differ from each other in diverse ways. Malacostraca is more clearly defined because all members of this group have abdominal appendages, an 8-segmented thorax, a gastric mill, and an abdomen of 6 (sometimes 7 or 8) segments. Entomostraca, there are four orders called Branchiopoda, Ostracoda, Copecoda, and Cirripedia, while Malacostraca there are three orders, namely: Isopoda, Stomatopoda, and Decapoda Order.

1. Enormostraca (small crustaceans)

   Characteristic: small in size, and is a lot of zooplankton found in sea water or fresh water. Its members consist of the Copepoda Order, the Cladocera Order, the Ostracoda Order, and the Amphipoda Order.

2. Malacostraca (large crustaceans)

   Its members consist of Isopoda Order (legged uniform) that used to live at sea, fresh water or land. For example, grasshopper shrimp and Decapoda Order (ten-legged) which has 5 pairs of limbs in the chest segment as legs, kinds like shrimp, crabs.

The malacostracans are the largest class of Crustacea, with more than 21,000 species including most of the larger and familiar forms such as prawns, crabs and lobsters. The body and appendages are specialised into thoracic and abdominal regions and head bears sensory and feeding appendages. The anterior thoracic limbs are commonly modified as additional mouthparts and limbs for capturing food, and are often chelate; the more posterior thoracic limbs are locomotory *pereiopods*. The abdominal appendages or pleopods are less sturdy and typically are used in slow swimming, the production of respiratory currents and in females for carrying eggs. The telson and uropods form a tail fan which enables rapid backward swimming when the abdomen is flexed – a behavior used to escape predators [2].

Crustacean development can be direct, in which the egg hatches into a fully formed but miniature version of the adult (as in most of the superorder Peracarida), or entirely anamorphic, in which change between successive molts consists essentially of increasing body size, adding segments and limbs, and developing existing limbs. Usually there is some metamorphosis, and at times this can be striking [7]. Typically, crustacean larvae have been grouped broadly into three main types, which are identified by the appendages primarily responsible for swimming; a 'nauplius' swims with its cephalic appendages, a zoea with its thoracic appendages, and a 'megalopa' with its abdominal appendages. The nauplius — zoea — megalopa' series represents a generalized developmental sequence as well, although most crustaceans do not pass through all three phases. Ecologically, the nauplius and zoea are usually dispersive phases and the megalopa is the transitional setlement phase.

4. The numerous economic benefits of Crustaceans to humans

Crustaceans have numerous direct and indirect benefits for the economy as well as human health. For instance, shrimps, crabs, lobsters and other large crustaceans are globally recognized as edible aquatic organisms. Furthermore, the Indonesian maritime has a yearly economic potential of 1.33 trillion USD. Shrimps are the most significant aquatic export commodity, and compose 45% of the country's total fishery export. The worldwide demand for Indonesian shrimp is approximately 560,000–570,000 tons per year, and about 57% of this figure is imported by the United States, the largest destination. Over 60% of the total aquatic produce exported to the US in 2016 was solely shrimp, and this was estimated to cost more than 1 billion USD, and to increase by 2017 in order to meet the increasing global demand. Indonesian shrimp is often exported while frozen or after removing the heads and shells [8]. Meanwhile, copepods, water insects, krill and other small zooplanktonic crustaceans, connect the food chain between photosynthetic phytoplanktons and larger carnivores, including whales and fishes. These petite crustaceans (zooplanktons) are therefore staple nourishment for large aquatic organisms, as several larger vertebrates consume spiders and insects.

4.1 Aesthetic value

The crustacean subphylum is considered unique, magnificent and delectable by humans of all ages. These arthropods exist in several shapes, sizes, colors and forms, and are astounding creations of nature. In addition, small crustaceans, including spider, ghost, fiddler (Ocypodidae), moon (Matutidae), rock, lightfooted, paddler (Graspidae), arrow, stone (Leusiidae), box (Calappidae), coconut, and hermit crabs, fairy, cherry, ghost, bamboo and, mantis banana shrimps, as well as freshwater, blue, and tiger crawfishes, as well as blue lobsters, are highly fascinating and invigorating, and therefore often used in aquatic exhibitions. Some prevalent interesting species include freshwater Atyid shrimps from the genus *Caridina, Palaemonetes*, *Atyopsis*, *Triops*, as well as *Neocaridina*, and crawfishes of the genus *Cambarellus* and *Procambarus.*

4.2 Commercial value

The cultivation of crustaceans is of great significance to the global aquaculture industry, as the arthropods are rich in protein, and possibly help to meet the food requirements for mankind's ever increasing population. Marine shrimps, crabs, prawns, and lobsters are valuable food sources, and therefore of substantial economic importance to aquatic industries around the world. Furthermore, crustacean aquaculture produce, particularly the true lobsters (*Panulirus versicolor, Homarus gammarus, and Homarus americanus*), are costlier compared to other sources of animal protein. The class of crustacean often cultivated for consumption by humans is Malacostraca, while crabs account for 20% of the marine crustacean species captured, reared and used worldwide. This amounts to about 1.5 million tons per year, and the specie *Portunus pelagicus* comprises one-fifth of this total. Lobster are a magnificent worldwide delicacy and lobster fishing, often referred to as *lobstering,* is the act of collecting marine or spiny lobsters, as well as crawfish, for commercial purposes. The commercial cultivation of shrimp was first practiced in the 1970's, and production has now developed steeply. Currently, about 75% of the shrimp and prawn cultivated in the world are produced in Asia (particularly, Indonesia, China, and Thailand). These are majorly genus penaeid, and two species, the Giant tiger prawn (*Penaeus monodon*) and the Pacific white shrimp (*Litopenaeus vannamei*, formerly *Penaeus vannamei*) comprise about 80% of all cultivated shrimp [9].

Meanwhile, *Macrobrachium* is the only cultivated genus of freshwater prawns. The species *M. nipponense*, *M. rosenbergii*, and *M. malacolmsonii* are majorly cultivated within the aquaculture industry. In addition, ornamental shrimps and crawfishes are often reared in South East Asia. The commercial fishing of Krill, marine crustaceans of the order Euphausiacea, closely resembling shrimp, as food for people and domesticated animals was first practiced during 19th century, and probably even earlier in Japan. These small invertebrates are present in oceans all around world, including the Southern Ocean, as well as the water body surrounding Japan. About 15o,000 to 20o,000 tons (minimum of 150,000–200,000 and maximum of 170,000–220,000) of krill are captured each year, and this is mostly obtained from the Scotia Ocean. This capture is mainly utilized as feed for aquaculture and aquarium organisms, bait for sport fishing, or to produce pharmaceuticals. Krill is often consumed by people in Japan as well as Russia and is referred to as *okiami* in Japan. In addition, some copepods and branchiopods (fairy shrimp and clam shrimps) are reared commercially for use in fish-farms and aquariums. These organisms therefore provide employment opportunities and recreational interests, through stocking, picking, feeding, sorting, and other activities related to crustacean gathering or cultivation.

4.3 Gastronomic value

Archaeologists have shown shellfishes were first used as food over hundreds of thousands of years ago. Practically all the cuisines prepared around the world involve these organisms as a significant protein source, especially in countries around coastal regions. The term "shellfish" is both a fishery and culinary word for edible aquatic invertebrates with exoskeleton, including several species of crustaceans, molluscs and in some cases, echinoderms. Thus, crustaceans are significant in shellfish seafood. These organisms are mainly aquatic and often harvested from saltwater bodies, however, some live in freshwater. Meanwhile, *Cardisoma guanhumi* and some other terrestrial crabs are also consumed, especially in the Caribbeans. Numerous organisms, from smaller animals like penguins and fish to the larger seals and even baleen whales feed on krill. Krills convert nutrient from consumed prey into an appropriate form for consumption by larger animals unable to feed directly on the diminutive algae, and are therefore, a crucial aquatic food chain component. Some species including northern krill, possess a rather small filtering basket and usually prey on copepods as well as larger zooplankton. The most popular commercially relevant species of the Euphausiidae genus include the Antarctic (*Euphausia superba*), Pacific (*Euphausia pacifica*) and Northern krills (*Meganyctiphanes norvegica*). In addition, the most dominant zooplanktons, marine copepods and ostracods are the major food sources for whales, small fishes, seabirds and crustaceans, including krill, in marine, brackish or fresh water bodies [10]. Branchiopods (brine shrimp and fairy shrimp) are utilized as nourishment of fish fry/aquarium fish food. The 90 species of krill are marine, pelagic, shrimp-like animals with shallow carapace. e.g. *Euphausia, Meganyctiphanes*. Species frequently have wide longitudinal but confined latitudinal dispersions and may be exceedingly various, often forming dense feeding swarms. Euphasids range from 40 to 150 mm in length. They are major nourishment source, particularly in Southern Ocean, for predators such as fish and squid and large scale filter-feeders such as baleen whales, and they are of expanding financial significance to humans [2].

4.4 Biomedical value

The shells of crabs and other crustaceans are used in medicine to treat and prevent inflammatory diseases. Some researchers at Florida Atlantic University have developed an orally administered crustacean microparticle dietary supplement to prevent and treat IBD and other inflammatory diseases, using the shells of crabs and other crustaceans. These chitin or chitosan microparticles undergo anti-inflammatory mechanisms applicable in the development of novel preventive and therapeutic substances for treating *inflammatory bowel disease* (IBD). The shells are a major, readily available waste produced by the seafood industry, and therefore an appropriate alternative for expensive and ineffective pharmaceutics, while krill oil is used as a dietary supplement. Two articles have been published with regard to the clinical applications of this oil in lipid lowering, arthritis pain relief and function, as well as C-reactive protein. The three most active medicinal components of krill oil are, the fish oil-like omega-3 fatty acids, the omega-3 fatty acids joined with phospholipids through conjugation, majorly phosphatidylcholine (marine lecithin), and the antioxidant, astaxanthin. Meanwhile, a study by the McGill University juxtaposed the impacts of fish and krill oils on cholesterol levels. The results showed a krill oil dosage of 1–3 g per day results in a more optimal treatment for hyperlipidemia, compared to a similar dose of fish oil.

4.5 Bioindicator and biomonitor

Crustaceans usually serve as bioindicators or biomonitors in various aquatic environment settings. This is because the arthropods are rather successful, and found in numerous environments including terrestrial, brackish, marine and freshwater habitats. The creatures are therefore, the perfect subjects for comparative analyses. Furthermore, some characteristics peculiar to crustaceans, particularly reproduction schemes, are possibly significant for interpreting data obtained from bio-indication studies based on these organisms, as well as in advancing focus on ecotoxicology. Thus, this presentation aims to highlight the use crustaceans, as biomonitors or bioindicators, especially in freshwater bodies. These two terms have currently not been distinguished. The term "bioindicator" refers to the characterization of a group of organisms in a particular field (in terms of statistics), in order to obtain information regarding the habitat, using the organisms' presence or absence, life history, or population (based on abundance, age distribution age, genetic composition or conditional index), as the study variables. Meanwhile, bio-monitoring involves characterizing living things in a bid to determine the bioavailability and geographical distribution of pollutants by measuring the concentration of accumulated chemicals within specific tissues or in the organism's entire body [11]. Planktonic copepods are crucial in global ecology and to the carbon cycle. Several scientists regard these organisms as the world's largest animal biomass, while others consider this to be the Antarctic krill (*Euphausia superba*). The *Calanus glacialis* inhabit the Arctic region's edge, and comprise up to 80% of the world's zooplankton biomass. In addition, some native species of crustaceans serve as bioindicators of pollution in freshwater bodies. *Palaemonetes argentinus* are able to function as proof of *environmental degradation* caused by pollutants in freshwater *ecosystems* [12]. Therefore, knowledge of pollutant bioaccumulation is required to completely comprehend the effect of pollution on aquatic ecology. Also, feeding and growth patterns, as well as migration and other animal behavior (for instance, precopula) are significant bio-indicators (or biomarkers).

4.6 Geological value

The most common arthropods in fossil records are Ostracods (seed shrimp). These fossils were first discovered around the Cambrian era, and continue to be found by archeologists, even in present day. M. B. Hart compiled a microfaunal zone layout base on Ostracoda and Foraminifera arthropods. Freshwater ostracods from the Baltic amber of Eocene era assumed to have been washed onto trees during surges, have also been discovered. These organisms are therefore of geological significance, particularly with respect to local or regional marine strata biozonation. The arthropods are also useful paleo-habitat indicators due to the prevalence, minute size, and the easily preserved generally-moulted and calcified bivalve carapaces, a commonly discovered microfossil [4].

4.7 Miscellaneous uses

Living copepods are used as feed in saltwater aquariums, and are generally useful for breeding marine species in captivity, particularly in reef tanks containing the notorious scooter blenny or the mandarin dragonet. Copepods are generally scavengers and in some cases, feed on coralline and other algae. These organisms, and other crustaceans, are popularly supplied as bait within the refugium, in saltwater aquariums. The mole crab or sand flea (*Emerita talpoida*), is a popular bait for game-fishing in oceans. Woodlice are terrestrial crustaceans having segmented, rigid, and long exoskeleton with fourteen jointed appendages, belonging to sub-order Oniscidea, and the order Isopoda. Over 3,000 known species belong to this order, and these organisms are mainly used in gardens for compost production and soil overturning.

4.8 Demerits or biodeterioration

4.8.1 Pests

Numerous water insects, especially spiny water fleas, *Bythotrephes & Cercopagis* species, are regarded as pests within fish farms. These fleas compete with juvenile fish for smaller zooplanktons, and because of quick reproduction rate, monopolize the food source, leaving the fish with lesser nourishment.

4.8.2 Spiny water-flea

These organisms are possible prey for fish, however, the presence of spine makes small fishes unable to swallow spiny water fleas. Studies have shown these organisms influence the survival and growth of young fish unfavorably, because of competition for food. Experts also assume the fishhook water flea to have similar impact. In addition, lake anglers most probably discover several hundred water fleas, resembling damp cotton, on fishing lines, as the long and spiny tails are easily entangled, and this poses a problem for anglers due to the clogging of the fishing rod's first line guide. Consequently, these fishermen are forced to resort to cutting the lines. Furthermore, woodlice feed on cultivated plants, including delicate seedlings and maturing strawberries, and also attack homes in groups looking for moisture, and therefore serve as an indicator for inadequate or excessive moisture. However, these organisms are not regarded as a serious household vermin, because woodlice are not disease vectors or destructive creatures. Conversely, Isopods, commonly referred to as gribble worms, belong to the

Limnoridae and Cheluridae families, and are productive borers, responsible for devastation of submerged timbers.

4.9 Poisonous Crustaceans

The species *Speleonectes atlantida*, an eyeless crustacean, of the Nectiopoda order was first discovered in August 2009 inside the Tunnel de la Atlantida, the longest submarine magma tube on Lanzarote, within in the west coast of the North African Canary Islands. This species possesses prehensile appendages, as well as venomous teeth serving as hypodermic needles. Meanwhile, the Coconut crab is consumed by indigenes of the Tuamotu Archipelago, and several cases of acute and lethal poisoning have been recorded within the region. Also, paralytic shellfish cause (saxitoxin) or tetrodotoxin poisoning, depending on the quantity of poison ingested.

4.9.1 Parasitism

Crustaceans are externally or internally parasitic to some organisms, especially as external parasites in fish. Furthermore, these arthropods are intermediate host for various pathogens, and first intermediate hosts for proceroid phase in several cestodes. *Argulus,* often referred to as "fish lice", and several other copepods, are particularly common parasitic crustaceans prevailing. These vampiric fish lice with dome-shells and beady-eyes, attach to scales using antennae in the shape of huge, spiked suckers.

Meanwhile, isopods, including woodlice and pillbugs belong to the family Gnathiidae, and are extremely similar to ticks, but lie in wait for fish. Prior to moulting into a non-parasitic phase, the fish-chigger feeds on blood for some days, and this cycle is repeated a few times during the organism's life cycle, while the *Cymothoa exigua*, known as tongue-eating lice, are possible parasitic crustaceans, belonging to the Cymothoidae family. These parasites invade fish from the gills, and subsequently, attach to the base of the tongue. In addition, the soft, tiny and globular pea-crabs (Pinnotherid) invade hosts durings as hatchlings, and subsequently feed and grow. The main hosts of parasitic crustaceans include oysters and other mussels, as well as echinoderm body cavities (sea cucumbers and urchins), snail shells, certain worms or and even sea squirts (within the gills).

4.9.2 Pea-crab in bivalve molluscs

Sacculina, and other rhizocephalans ("root heads") are born as "cyprid", as observed in several barnacles, then swim freely in search of host, unlike other barnacles. These hosts include other crustaceans, including live crabs, and in some cases, about half of the entire crab population is infested. Some other parasitic crustaceans include sarcotacids (skin-bags), pentastomids (five-mouths/tongue-worm), and siphonostomatoids (flesh-Anchors).

5. Conclusions

Crustaceans are a major group belonging to the Arthropoda subphylum, often used as food for animals and humans. Furthermore, at least 50,000 to 75,000 species of crustaceans have been recorded worldwide. These creatures, particularly crabs, shrimps, prawns, lobsters, and other decapods, are also the most significant source of animal protein for the human population, and are a significant sustainable part of modern culture and commerce, bio-medicine and economic development, aesthetics, gastronomy, geology, and many other fields.

Acknowledgements

Author is great full to Dr. Suripto Dwi Yuwono, the Dean of Faculty Mathematics and Natural Sciences, University of Lampung, Indonesia encouragement guidance in successful completion of this chapter. I wish to cordially thank my wife who has always encouraged and supported me on my way towards finishing the chapter.

Conflict of interest

Authors declare no conflict of interest.

Author details

Gregorius Nugroho Susanto
Department of Biology, Faculty of Mathematics and Natural Sciences, University of Lampung, Bandar Lampung, Indonesia

*Address all correspondence to: gnugrohos@gmail.com

References

[1] Anger K. The Biology of Decapod Crustacean Larvae. Biologische Anstalt Helgoland, Stiftung Alfred-Wegener-Institut für Polar- und Meeresforschung, 27498 Helgoland, Germany. 2001. https://www.researchgate.net/publication/266673708

[2] Greenaway P. The Crustacea. In: D.T. Anderson (Ed.), Invertebrate Zoology. Melbourne, Oxford University Press; 1998. p 286-318. https://www.researchgate.net/publication/277181821

[3] Kämpf J, Chapman P. Upwelling Systems of the World, Springer International Publishing Switzerland. 2016. DOI: 10.1007/978-3-319-42524-5_1

[4] De Grave S, Pentcheff N D, Ahyong S T, Chan T-Y, Crandall K A, Dworschak P C, Felder D L, Feldmann R M, Fransen C H J M, Goulding L Y D, Lemaitre R, Low, Martin J W, Peter K L Ng, Schweitzer C F, Tan S H, Tshudy D, Wetzer R. A Classification of Living and Fossil Genera of Decapod Crustaceans. Raffles Bulletin of Zoology, Supplement 2009. No. 21: 1-109.

[5] Lumenta C. Aquatic Invertebrates. Unsrat Press, Manado, Indonesia. 2017.

[6] Brusca R C, Brusca G J. Invertebrates. Sinauer Associates, Sunderland, MA. 1990.

[7] Harvey A W, Martin J W, Wet R. Phylum Arthropoda: Crustacea (Chapter 17) in Atlas of Marine Invertebrate Larvae. Academic Press. 2002. p 337-369.

[8] BKPM (Indonesia Investment Coordinating Board). Investing in Indonesia's Fisheries Sector: An overview of opportunities, capabilities and provisions. The EU-Indonesia Trade Cooperation Facility project, Jakarta, Indonesia. 2016.

[9] Pratiwi R. Economic aspects of shrimp biology are important. Oseana. 2008; 33 (2):15-24.

[10] Conway D V P. Marine zooplankton of southern Britain. Part 2: Arachnida, Pycnogonida, Cladocera, Facetotecta, Cirripedia and Copepoda (ed. A.W.G. John). Occasional Publications. Marine Biological Association of the United Kingdom, No 26 Plymouth, United Kingdom. 2012. 163 pp. DOI: 10.13140/2.1.4704.4800

[11] Rinderhagen M, Ritterhoff J, Zauke G P. Crustaceans as Bioindicators. in Biomonitoring of Polluted Water – Reviews on Actual Topics (A. Gerhardt, ed.), TransTech Publications – Scitech Publications, Environmental Research Forum. 2000; 9:161-194.

[12] Bertrand L, Monferrán M V, Mouneyrac C, Amé M V. Native crustacean species as a bioindicator of freshwater ecosystem pollution: A multivariate and integrative study of multi-biomarker response in active river monitoring. Chemosphere. 2018; 206:265-277. DOI: 10.1016/j.chemosphere.2018.05.002

Chapter 10

Lecanicillium spp. for the Management of Aphids, Whiteflies, Thrips, Scales and Mealy Bugs: Review

Sajjalavarahalli Gangireddy Eswara Reddy

Abstract

Lecanicillium spp. are potential microbial bio-control agent mainly used for the management of sucking insect pests such as aphids, whiteflies, scales, mealy bugs etc. and gaining much importance at present for management of pests. Due to indiscriminate use of chemical pesticides which results in development of resistance, resurgence, outbreak of pests and residue problem, the farmers/growers are forced to use biopesticides for sustainable agriculture. *Lecanicillium* spp. is promising biocontrol agent against sucking insect pests and can be used as one of the components in integrated pest management (IPM). However, optimum temperature and relative humidity are the major environmental factors, for the performance of *Lecanicillium* spp. under protected/field conditions. The present review is mainly focused on nomenclature of *Lecanicillium* spp., mode of infection, natural occurrence, influence of temperature and humidity on the growth, factors influencing the efficacy, virulence/pathogenicity to target pests, substrates used for mass production, safety to non-target organisms, compatibility with agrochemicals and commercially available products. This review is mainly useful for the researchers/students to plan their future work on *Lecanicillium* spp.

Keywords: entomopathogenic fungus, aphid, whitefly, virulence, mass production, safety

1. Introduction

The increased use of conventional chemical pesticides over the years has not only contributed to an increase in food production, but also has resulted in adverse effects on the environment and non-target organisms. In view of these side effects, the necessity for sustainable crop production through ecofriendly pest management technique is being largely felt in the recent times. Few biopesticides are available in the market, among them *Lecanicillium* spp. based microbial bio-pesticide gaining much importance for sucking pests for organic and sustainable agriculture [1–4]. Myco pesticides are potential microbial alternative to chemical pesticides and offer a number of benefits such as facility of growth on a variety of substrates, high virulence, trans cuticular penetration, broad host range, less expensive, safe to humans, animals and the environment. Therefore, this review is prepared by compiling the research work done on *Lecanicillium* spp. by various research groups on various aspects viz., nomenclature, mode of infection, natural occurrence, effect of temperature and humidity for the growth, factors influencing its efficacy, virulence and pathogenicity against target pests under laboratory/greenhouse/field, substrates used for mass production, safety to non-target organisms, compatibility with agrochemicals and commercially available products were discussed and presented.

2. Nomenclature of *Lecanicillium* spp.

The genus *Verticillium* contains diverse host ranges including arthropods, nematodes, plants and fungi [5]. The genus has been redefined using rDNA sequencing, grouping insect pathogens into the new genus *Lecanicillium* which includes *L. attenuatum, L. lecanii, L. longisporum, L. muscarium* and *L. nodulosum*, which were all formerly classified as *V. lecanii* [5–7].

3. Mode of infection

When *L. lecanii* conidia comes in contact with the host integument, it gets adhere to the epicuticle and germinate. Germinated conidia form germ tubes that penetrate cuticle directly or grow over the surface of the epicuticle. The germ tube penetrates by lysing both the epicuticle and the procuticle [8, 9]. This is accomplished by the mechanical pressure exerted by appresorium (penetration peg) and secretion of enzymes viz., proteases, chitinases and esterase's which plays an important role during cuticle penetration of insect host and also serve as cuticle degrading enzymes. The fungus proliferates throughout the insect's body, draining the insect of nutrients, and eventually killing it in around 48–72 hours. The mycotoxins produced by *L. lecanii* are bassianolide [10, 11], vertilecanin-A1, decenedioic acid and 10-hydroxy-8-decenoic acid) [12–14]. As the host nutrients are depleted, the blastopores' differentiate into elongated hyphae which extend outward from the body forming a mycelial mat of conidiophores over the surface of the integument resulting in mummification. Under favourable environmental condition, conidiophores mature giving rise to conidia which continues the disease cycle further.

4. R & D publications on different aspects of *Lecanicillium* spp.

The number of publications related to *Lecanicillium* spp. from 1971 to 2020 was presented in the **Figures 1** and **2**. The data clearly indicated that, during 1971–80^s

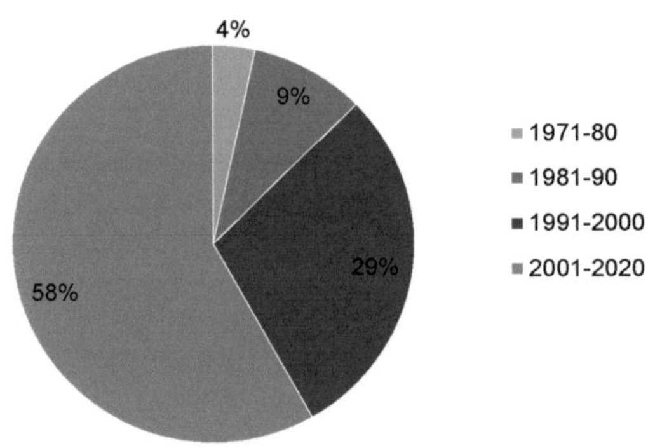

Figure 1.
Per cent R & D publications related to Lecanicillium *spp.*

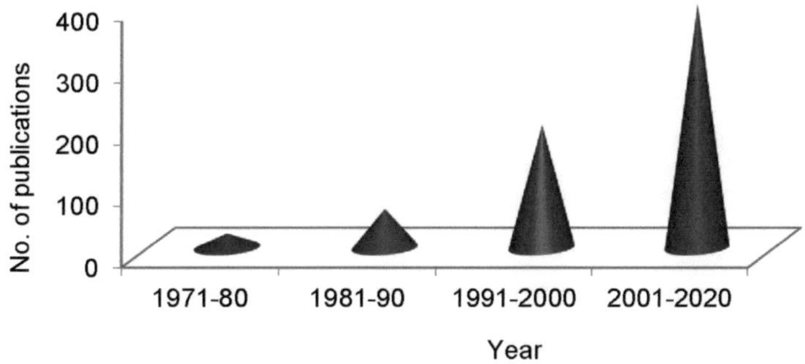

Figure 2.
R & D publications on Lecanicillium *spp.*

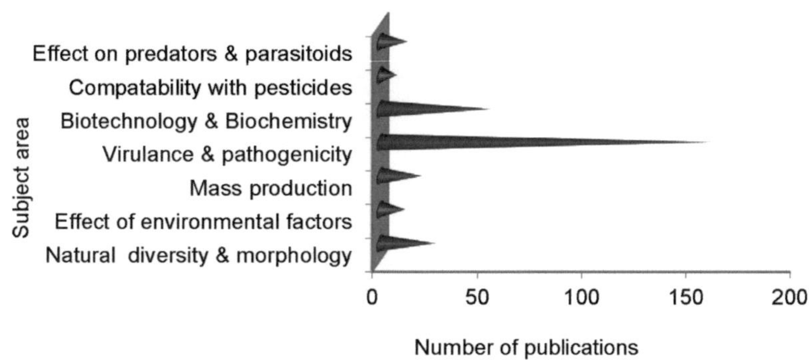

Figure 3.
R & D publications on various aspects of Lecanicillium *spp.*

the publications were completely nil, but during 1981–91^{s}, the R&D work has been initiated in the entire world and the publications were increased gradually reaching 58% during 2001–2020 (**Figure 2**). While, considering the number publications on various aspects of *Lecanicillium* spp., more research work has been done on virulence and pathogenicity (**Figure 3**) followed by biotechnology and biochemistry as compared to morphology, diversity, ecology, mass production. The number of publications was meagre on effect of environmental factors (temperature and humidity), safety to natural enemies and compatibility with pesticides [15].

5. Natural occurrence of *Lecanicillium* spp.

Lecanicillium spp. is the most widely distributed and generally found on infected insects both in temperate and tropical areas throughout the world. There are number of reports on natural infection of *Lecanicillium* spp. on different insect pests but out of the reported insects and pests, maximum are sucking pests belonging to Hemiptera, Thysanoptera and Acarina which indicates its possible spectrum for use as a biocontrol agent for pest management. Reports of natural occurrence of *Lecanicillium* spp. on sucking insects presented in the **Table 1**.

Strain/isolate	Host	Location	Reference
L. lecanii (Is-2, Is-5)	*M. persicae*	Israel	[14]
L. lecanii (Is-6)	*Acrithosiphon pisum*	Israel	[14]
L. lecanii (R-1)	*T. vaporariorum*	Russia	[14]

Strain/isolate	Host	Location	Reference
L. lecanii (Vl6063)	*T. vaporariorum*	Halifax, Canada	[2]
L. lecanii (V0175)	*B. tabaci*	Guangdong, China	[2]
L. lecanii (Vp28)	*Pseudococcus* sp.	Guangdong, China	[2]
L. lecanii (ICAL4)	*Nasonovia ribisnigri* in lettuce	Madrid	[16]
L. lecanii (ICAL6)	*M. persicae* in pepper	Madrid	[16]
L. lecanii (41185)	*M. persicae*, *T. vaporariorum*	Korea	[17]
L. longisporum (6541)	*Aphis gossypii*	UK	[17]
L. longisporum (6543)	*M. persicae*	UK	[17]
L. longisporum (4078)	*M. persicae*	Denmark	[17]
L. longisporum (HRI 1.72)	*Macrosiphoniella sanbornii*	UK	[18]
L. lecanii (ARSEF 7207)	*T. vaporariorum*	Argentina	[16]
L. longisporum (ARSEF 7461)	*T. vaporariorum*	Argentina	[16]
L. muscarium (ARSEF 7460)	*T. vaporariorum*	Argentina	[16]
L. lecani (ICAL3)	*Macrosiphum euphorbiae* in tomato	Madrid, Spain	[19]
L. lecanii (ITEM 3757)	*Brevicorne brassicae* in Cabbage	Bari, Italy	[20]
L. lecanii	*S. bispinosus* on tea	Tamil Nadu, India	[21]
L. sabanense sp. nov	*Pulvinaria caballeroramose*	Bogota (Columbia)	[22]
L. attenuatum ZJLSP07 and *L. psalliotae* ZJLA08	*Diaphorina citri*	Taizhou (Zhejiang Province, China	[23]
L. lecanii (FI 2482) and *L. muscarium* (FI 2481)	*Thaumastocoris peregrinus*	South-East Uruguay	[24]

Table 1.
Natural occurrence of Lecanicillium spp. on different sucking insect pests.

6. Effect of temperature and humidity on the growth of *Lecanicillium* spp.

Temperature and humidity are the main factors influencing the growth of the fungus. Effect of different temperature on conidial germination, growth rate, colony size and mycelial growth was discussed and presented in **Table 2**.

Temperature	5°C	10°C	15°C	20°C	25°C	30°C
Water activity (aw)	—	0.985	0.99	0.98	0.975	[25]
Strain/isolate	% Conidial germination					
L. longisporum (Vertalec)	—	—	98	98	28.7	—
L. muscarium (Mycotal)	—	20.6	98	98	98	—
PFC 1	—	—	—	50.6	47	—
PFC 3	—	—	—	49.7	86.6	—

Temperature	**5 °C**	**10°C**	**15°C**	**20°C**	**25°C**	**30°C**	
PFC 10	—	64.7	47.7	14.7	—	—	
PFC 11	—	—	98	98	49.3	—	
PFC 13	—	88	98	98	98	—	
			Mean radial growth rates (mm/day)				
Strain/isolate	5 °C	10 °C	15 °C	20 °C	25 °C	30 °C	[25]
L. longisporum (Vertalec)	0.21	0.66	1.10	1.31	1.86	0.55	
L. muscarium (Mycotal)	0.22	0.59	1.03	1.59	2.03	0.59	
PFC 1	0.16	0.43	0.90	1.13	1.64	0.69	
PFC 3	0.15	0.54	1.03	1.35	1.86	0.83	
PFC 10	0.18	0.63	1.02	1.40	2.07	0.05	
PFC 11	0.17	0.58	1.03	1.25	2.05	0.05	
			Mean colony size (diameter: mm)				
Strain/isolate	5 °C	10 °C	15 °C	20 °C	25 °C	30 °C	[26]
Vertalec	5.0	18.6	34.1	50.2	52.1	5.0	
Mycotal	12.1	20.5	31.5	42.1	47.3	8.3	
B-2	11.6	21.3	25.4	46.2	53.6	26.9	

Table 2.
Effect of temperature on growth of Lecanicillium spp.

6.1 Temperature

Temperature affects the *Lecanicillium* spp. in different ways by influencing the germination, growth and viability of the fungus in the host insect and environment. High temperature inactivates the fungus before contact with the pest insect or may reduce or accelerate the growth within an insect depending on the temperature requirements of the fungus and the host insect. In contrast, low temperatures reduce or stop the germination and growth. Optimal germination and growth rates of *Lecanicillium* spp. range between 23°C and 28°C, growth rapidly slows >30°C and ceases at 34 to 37°C. Similarly, conidial germination is adversely affected by temperatures above 30°C. Temperature below 16°C increasingly slows germination and growth and thus affects efficacy in terms of a longer survival of the target population. *Lecanicillium* strains showed optimum growth at 25°C; the aerial conidia of *Lecanicillium* strains germinate in a broad temperature range (15–30°C) and *L. lecanii* 41,185 was the only strain with conidial germination at 35°C [27].

Effect of different temperature on conidial germination, growth rate, colony size and mycelial growth of *L lecanii* was discussed and presented in **Table 2**. At 25°C and 0.975 a_w (water activity) conidial germination occurred in all the isolates ranging from 28.7 to 98% whereas isolate PFC 10 no conidial germination had. Per cent germination decreased from highest values at 25°C to the lowest trend at 10°C in Mycotol (20.6°C). Maximum germination of conidia was observed between 15 and 25°C [25]. Most of the isolates showed growth at 5 and 30°C and mean growth rate increased as temperature increased. Optimum growth rate occurred at 25°C (1.64 to 2.07 mm) for all isolates) [25]. Colony size of the fungus was influenced by the temperature, the colony growth is maximum at 25°C (47.3 to 53.6 mm) as compared to the temperature between 5 to 20°C [26]. The optimum temperature for the mycelial growth of *L. lacanii* CA-1-G was 23°C (37.57 mg/cm^2) and 26°C (39.43 mg/cm^2) as compared to 20°C (29.43 mg/cm^2), 29°C (20.7 mg/cm^2) and 32°C (20.63 mg/cm^2). Similarly, *L. lecanii* grew and sporulated over a wide range of temperatures (20–32°C). The optimum temperature for growth was 23°C (46.45 x10^5 conidia cm^{-2}) or 26°C (33.76 x10^5 conidia cm^{-2}) for *L. lecanii* CA-1-G [28]. Virulence of *Lecanicillium* spp. isolates was evaluated against third instar *T. vaporariorum* on tomato plants at 23°C. Colony radial growth, conidial production and germination decreased with the reduction in water activity, while 32°C was extremely detrimental for all fungal isolates. However, some isolates were able to grow and produce conidia at low water activity and high temperature [29]. *L. muscarium* can multiplied in temperature range of 15–30°C but optimum temperature against *M. persicae* between 20 to 30°C [30].

6.2 Humidity

Humidity is another important environmental factor affecting the efficacy and survival of *Lecanicillium*. Spore germination on the insect cuticle and sporulation after outgrowth of the dead host insect require high moisture. Generally high humidity is required for germination of spores under in vitro, insects can become infected at much lower humidity. Under fluctuating humidity, daily saturated humidity requirement of at least 16 h for causing death in *Trialeurodes vaporariorum* (Westwood) infected with *L. lecanii* [31]. Several previous studies provided evidence that a threshold time period at high humidity was required for infection. Conidia of *L. lecanii* required at least 72 h at 100% RH and 20°C before removal to 70% RH to reach >90% infectivity of *Myzus persicae* (Sulzer) [32]. Similarly, at 25°C temperature and 75% relative humidity (RH), *L. lecanii* 41,185 showed highly virulent pathogenicity (100%) against *M. persicae* and *Aphis gossypii* Glover [27]. Application of *L. longisporum* against *A. gossypii* on cucumber in controlled environment (Temperature; 19–26°C and humidity; 80–98%) resulted in 100% mortality [32, 33]. *L. muscarium* grow at optimum temperature but higher mortality observed against *M. persicae* between 55 and 90% humidity [30].

7. Factors influencing the efficacy of *Lecanicillium* spp. against sucking insect pests

The virulence and pathogenicity of *Lecanicillium* spp. vary with strain, stage of the insect and dose of the fungus.

7.1 Strains

Virulence of the *Lecanicillium* spp. varies with strain to strain or isolate to isolate. The isolate ICAL6 was more virulent (LC_{50} = 1.05 x 10^7 conidia mL^{-1}) to nymphs of *M. persicae* than *Macrosiphum euphorbiae* (Thomas) (LC_{50} = 1.26 x 10^7 conidia mL^{-1})) and *Nasonovia ribisingri* (Mosley) (LC_{50} = 2.78 x 10^7 conidia mL^{-1}) [19]. The strain Vl 6063 imported from Canada was more virulent to *Bemisia tabaci* (Gennadius) (2.57 x 10^5 conidia mL^{-1})) than the domestic strains V3450 and Vp28 (LC_{50} = 6.03 x 10^5 conidia mL^{-1}) [2]. *L. lecanii* @ 1x10^7 conidia mL^{-1} is more effective against nymphs of *Plannococcus citri* (84% mortality) after six days of treatment as compared to *L. longisporum* (59% mortality) [34]. *L. muscarium* isolate FI 2481 @ 1x10^7 conidia mL^{-1} was more effective against *Thaumastocoris peregrinus* (72% mortality) as compared to *L. lecanii* isolate FI 2482 which reported 50% mortality [24]. Similarly, *L. lecanii* hybrid strain 2aF4 was more promising (LC_{50} = 5.3x10^4 conidia mL^{-1}) for the management of *Trialeurodes vaporariorum* than *L. lecanii* 2aF4 (LC_{50} = 7.8x10^4 conidia mL^{-1}) [35].

7.2 Stage of the insect

Stage of host plays important role in the success of *Lecanicillium* spp. and not all stages of insect life cycle are equally susceptible to fungal infection. So, the fungal application can be successful against the particular pest when it can be done at the condition where the susceptible stage or weaker stages of the particular pest become dominant among population.

First and third instar nymphs of *B. tabaci* (38 and 65% mortality) were significantly more susceptible to *L. muscarium* than the fourth instar (15%) in verbena plants. Similarly, first and second instars *B. tabaci* was more susceptible (50 and 55% mortality) than the third and fourth instar (25 and 20% mortality) on tomato foliage [36]. *L. lecanii* (ARSEF 7460) showed higher mortality against nymphs of *T. vaporariorum* followed by *L. longisporum* (ARSEF 7207) and *L. muscarium* (ARSEF 74601) @ 1 x 10^7 conidia mL^{-1}) [16]. The pathogenicity of *L. lecanii* strains was more in pupae (59–72.5%) than adults (34–52.6%) after 6 days of inoculation [14]. *L. lecanii* (2.8 x 10^7 conidia/ml) isolated from *Scirtothrips bispinosus* (Bagnall) in tea showed higher mortality against larvae (60%) than adults (30%) of *S. bispinosus* under laboratory at same dose [21]. Mortality of nymphs of *Plannococcus citri* were more susceptible (84% mortality) after six days of treatment to *L. lecanii* @ 1x10^7 conidia mL^{-1} as compared to adults which showed 40% mortality [34]. *L. lecanii* hybrid strain 2aF43 @ 1x10^7 conidia mL^{-1} showed more efficacy against first instar nymphs of *T. vaporariorum* (68% mortality) as compared to 4th instar nymphs (30% mortality) and adults (60% mortality). Similarly, *L. lecanii* hybrid strain 2aF4 is more effective against first instar nymphs (46% mortality) as compared to 4th instar nymphs (30% mortality) [35].

7.3 Dose/inoculums level

Fungal inoculum level is the important factor which affects the performance. It is general trend that the higher fungal inoculum level gives higher insect mortality. However, sufficient inoculum level should be worked out for the particular pest to prevent the over inoculum wastage and to achieve higher mortality. Higher dose of *L. lecanii* (1.2 x 10^9 conidia ha^{-1}) was caused 92.30 and 80.93% mortality of *Brevicoryne brassicae* Linnaeus and *Aleurodicus disperses* (Russell) respectively at 10 days after treatment in the laboratory, whereas in field conditions *L. lecanii* (Vl3) at 2 x 10^{12} conidia ha^{-1}) causing 61.16% and 66.50% mortality of *B. brassicae* and *A. craccivora* respectively [2].

8. Efficacy of *Lecanicillium* spp. against sucking pests under laboratory/greenhouse/field

Efficacy of *Lecanicillium* spp., against aphids, whiteflies, thrips, scales and mealy bugs in the laboratory/greenhouse/field conditions w.r.to its mortality, LC_{50} and LT_{50} values were presented in the **Table 3**.

Strain/isolate	Conditions (Lab, GH, F)	Pest	Mortality/LC_{50}/LT_{50}	Temperature (°C)	Humidity (%)	References
Lecanicillium lecanii	Lab	*Bemisia argentifolii*	95–98%	20–25	100	[14]
L. lecanii (HRI 1.72)	Lab	*A. fabae*	LT_{50} (2.79 d)	10–23		[23]
L. lecanii (HRI 1.72)	Lab	*M. persicae*	LT_{50} (3.39 d)	23		[37]
L. lecanii (VI6063)	Lab	*B. tabaci*	(94.9%) $LC_{50} = 2.57 \times 10^5$ Conidia mL^{-1}	25	95	[2]
L. lecanii (V3450)	Lab	*B. tabaci*	86.9 ($LC_{50} = 6.03 \times 10^5$ conidia mL^{-1})	25	95	[2]
L. longisporum		*M. persice, Macrosiphum euphorbiae, Aulacorthum solani*	(LT_{50} = 2.4;1.8; 2.0 d) 100% mortality	25	95	[38]
L. longisporum (HRI 1.72)	Lab	*M. persicae, A. fabae, Acrithosiphon pisum, Sitobion avenae*	LT_{50} = 74–78 h			[18]
L. longisporum	Cucumber	*A. gossypii*	100% (LT_{50} = 6.9 d)	25.8	80.6	[33]
L. longisporum or L. muscarium	Lab	*Frankliniella occidentalis*	95%	20	70%	[39]
L. lecanii	Lab	*A. craccivora*	($LC_{50} = 2.5 \times 10^4$ spores mL^{-1}) ($LT_{50} = 3.9 \times 10^8$ spores mL^{-1})			[40]
L. muscarium ($1x10^7$ spores/ml	Verbana, tomato (GH)	*B. tabaci*	65 and 55% mortality	20	95	[41]
L. muscarium ($1x10^7$ conidia/ml)	Verbana, tomato (GH)	*B. tabaci*	85 and 80%	20	85	[36]
L. longisporum	Cucumber (GH)	*A. gossypii*	100%	19.0	80.2	[42]
L. lecanii	Tea (F)	*S. bispinosus*	30–60%	—	—	[21]
L. attenuatum ZJLSP07 and *L. psalliotae* ZJLA08	Lab	*Diaphorina citri*	100% ($1x10^8$ conidia/ml)	25	90	[23]
L. attenuatum (SD-16, SDMP1 and 2)	Lab	*M. persicae*	100%	25	>90	[43]

Strain/isolate	Conditions (Lab, GH, F)	Pest	Mortality/LC_{50} /LT_{50}	Temperature (°C)	Humidity (%)	References
L. lecanii	Lab	*M. persicae, A. gossypii*	100% (1x10^8 conidia/ml)	20	>90	[44]
L. lecanii (JMC-01)	Lab	*B. tabaci*	82.2% (1x10^8 conidia/ml)	25	70	[45]
L. lecanii (FI 2482) and *L. muscarium* (FI 2481)	Lab	*Thaumastocoris peregrinus*	50 and 72%	25	65	[24]
L. lecanii 2aF4 3 and 2aF4	Lab	*T. vaporariorum*	83% (LC_{50} = 5.3x10^4 conidia/ml) and 84% (LC_{50} = 7.8 x10^4 conidia/ml)	23	99.6	[35]

GH; Green house, F; Field, LC_{50}; Lethal concentration to kill 50% insects, LT_{50}; Lethal time to kill 50% insects.

Table 3.
Efficacy of Lecanicillium spp. against sucking pests under laboratory/greenhouse/field.

9. Substrates used for mass production of *Lecanicillium* spp.

Lecanicillium spp. can be mass multiplied by solid state fermentation (SSF) and liquid state fermentation (LSF) using different growth media. In SSF, different grains, agars and non-synthetic solid media were used for mass production of *Lecanicillium* spp. (**Table 4**).

Substrates	Conidia/Spores	References
Media		
Sabaroud dextrose agar	2.87×10^7 conidia cm $^{-2}$	[46]
Malt extract agar	5.23×10^7 conidia cm $^{-2}$	
Nutrient agar	1.07×10^7 conidia cm $^{-2}$	
Corn meal agar	0.09×10^7 conidia cm $^{-2}$	
Yeast peptone dextrose agar	4.58×10^7 conidia cm $^{-2}$	
Potato dextrose agar	2.91×10^7 conidia cm $^{-2}$	
Grains		
Rice	8.43×10^8 spores g $^{-1}$	[47]
Wheat	9.13×10^8 spores g $^{-1}$	
Sorghum	11.31×10^8 spores g $^{-1}$	
Pearl millet	10.17×10^8 spores g $^{-1}$	
Finger millet	9.76×10^8 spores g $^{-1}$	
Maize	7.54×10^8 spores g $^{-1}$	
Rice	1.97×10^9 spores g $^{-1}$	[48]
Sorghum	1.90×10^9 spores g $^{-1}$	
Finger millet	1.66×10^9 spores g $^{-1}$	
Wheat	1.65×10^9 spores g $^{-1}$	
Corn	1.84×10^9 spores g $^{-1}$	
Polished rice	5.7×10^8 conidia g $^{-1}$	[49]
Cooked rice	1.5×10^9 conidia g $^{-1}$	
Rice bran	1.4×10^9 conidia g $^{-1}$	
Crushed bajra +1% yeast extract (YE)	17.49×10^8 conidia g $^{-1}$	[4]
Crushed sorghum +1% YE	10.34×10^8 conidia g $^{-1}$	
Crushed navane +1% YE	3.52×10^8 conidia g $^{-1}$	
Crushed maize +1% YE	4.80×10^8 conidia g $^{-1}$	
Crushed rice +1% YE	24.59×10^8 conidia g $^{-1}$	
Crushed wheat +1% YE	3.54×10^8 conidia g $^{-1}$	
Rice bran	24×10^7 conidia g $^{-1}$	[34]
Agro wastes		
Crushed maize cobs +10% molasses	10.07×10^4 conidia/g $^{-1}$	[4]
Wheat bran +10% molasses	18.76×10^4 conidia g $^{-1}$	
Rice bran +10% molasses	30.86×10^4 conidia g $^{-1}$	
Baggase +10% molasses	10.88×10^4 conidia g $^{-1}$	

Substrates	Conidia/Spores	References
Press mud +10% molasses	7.90 x 10^4 conidia g^{-1}	
Sugarcane molasses 3%	8.35 x 10^8 spores ml^{-1}	[48]
Sugarcane molasses 4%	8.56 x 10^8 spores ml^{-1}	
Sugarcane molasses 5%	8.42 x 10^8 spores ml^{-1}	
Non synthetic solid media		
Carrot	2.17 x 10^8 spores g^{-1}	[47]
Jack seeds	4.11 x 10^8 spores g^{-1}	
Ladies finger	3.12 x 10^8 spores g^{-1}	
Rice husk	1.27 x 10^8 spores g^{-1}	
Saw dust	0.69 x 10^8 spores g^{-1}	
Beet pulp	23 x 10^7 conidia g^{-1}	[34]
Non synthetic liquid media		
Coconut water	5.27 x 10^8 spores g^{-1}	[47]
Rice cooked water	2.11 x 10^8 spores g^{-1}	
Rice wash water	3.12 x 10^8 spores g^{-1}	
Wheat wash water	1.21 x 10^8 spores g^{-1}	
Liquid media		
Potato carrot broth	6.50 x 10^7spores mL^{-1}	[48]
Potato dextrose broth	3.95 x 10^7spores mL^{-1}	
Potato sucrose broth	6.30 x 10^7 spores mL^{-1}	
Jaggery yeast broth	2.45 x 10^7 spores mL^{-1}	
Sucrose yeast broth	2.50 x 10^7 spores mL^{-1}	
Molasses yeast broth	8.33 x 10^7 spores mL^{-1}	

Table 4.
Substrates (media, grains, agro wastes) used for mass production of Lecanicillium spp.

Among grains, rice is most suitable substrates for mass production (1.97 x 10^9 spores g^{-1}) followed by sorghum (1.90 x 10^9 spores g^{-1}) as compared to finger millet, wheat and corn (1.6–1.80 x 10^9 spores g^{-1}) [48]. Similarly, crushed rice +1% yeast extract recorded higher spore yield (24.59 x 10^8 conidia g^{-1}) followed by crushed bajra +1% yeast extract (17.49 x 10^8 spores g^{-1}) as compared to crushed sorghum, maize and wheat [4].

Among different agro wastes used for multiplication of *L. lecanii, the* growth and sporulation were found to be better on rice bran +10% molasses (30.86 x 10^4 conidia g^{-1}) followed by wheat bran +10% molasses (18.76 x 10^4 conidia g^{-1}) and rice bran (15.98 x 10^4 conidia g^{-1}). Complete inhibition of growth and reproduction of the fungus was noticed on bagasse and pressmud with 1 per cent yeast extract alone. However, growth was recorded when bagasse and press mud was supplemented with 10% molasses (10.88 and 7.90 conidia g^{-1} respectively) [4]. Among agars, malt extract agar (MEA) yields high conidia production (5.23 x 10^7 conidia cm^{-2}) followed yeast peptone dextrose agar (4.58 x10^7 conidia cm^{-2}) as compared to potato dextrose agar and sabaroud dextrose agar (2.91 and 2.87 x 10^7 conidia cm^{-2} respectively) [46]. In non-synthetic media, jack seeds produced high spore yield (4.11 x 10^8 spores g^{-1}) followed by ladies' finger (3.12 x 10^8 spores g^{-1}), carrot (2.17 x 10^8 spores g^{-1}) and rice husk (1.27 x 10^8 spores g^{-1}) [47].

In LSF, molasses yeast broth (MYB) supported maximum spore production of *L. lecanii* (8.33 x 10^7 spores ml^{-1}) followed by potato carrot broth (6.5 x 10^7 spores/ml) and potato sucrose broth (6.3 x 10^7 spores ml^{-1}) as compared to Sucrose yeast broth, Jaggery yeast broth and Potato dextrose broth (2.45–3.95 x 10^7 spores mL^{-1}). Among non-synthetic liquid media, coconut water produced higher spores (5.27 x 10^7 spore's g^{-1}) and biomass production than rice wash water (3.12 x 10^7 spore's g^{-1}) as compared to rice cooked water and wheat wash water (1.21–2.11 x 10^7 spores g^{-1}) [24]. The growth of *L. longisporum* conidial spores are higher in rice bran [24 x 10^7 conidia g^{-1}] as compared to beet pulp [23 x 10^7 conidia g^{-1}] [34].

10. Safety of *Lecanicillium* spp. to parasitoids/predators/pollinators

The safety of any bio control agent to parasitoids/predators/pollinators is the important aspect which should be studied thoroughly before its commercialization to avoid the hazards and disturbance of ecological balance. Effect of *L. lecanii* on aphid parasitoid *Aphidius colemani* (Viereck) which showed the normal development (approximately 90% adult emergence) when its cotton aphid, *A. gossypii* host was treated with *L. lecanii* conidia 5 or 7 days after parasitization. Fungus exposure 1 day before or up to 3 days after parasitization, however, reduced *Aphidius colemani* (Viereck) emergence from 0 to 10%. They suggested that the parasitoid and fungus may be used together for aphid bio control [50]. *L. lecanii* showed pathogenicity against predatory mite, *Phytoseiulus persimilis* Athias-Henriot but its effect was lower than that of spider mite, *Tetranychus urticae* (Koch) [51]. *L. lecanii* is safer to predatory coccinellid, *Coccinella septempunctata* Linnaeus and predatory mites, *Amblyseius ovalis* (Evans) and *Amblyseius longispinosus* (Evans) under field conditions [52]. The fungus *L. lecanii* was not pathogenic to *Chrysoperla carnea* (Stephens), *Coccinella septempunctata* (Linnaeus), *Episyrphus balteatus* (De Geer) and *Samia cynthia ricini* (Boisduval), but was found to be pathogenic to *Bombyx mori* (Linnaeus). Parasitism, adult emergence and adult longevity of *Trichogramma chilonis* (Ishii) were affected by fungal treatments. Aphid mummification and *Diaeretiella rapae* adult emergence were affected by the fungus. Results suggest that *L. lecanii* is compatible with natural enemies of cabbage aphid, *T. chilonis* and is harmless to silk worm [53]. *L. muscarium* at 10^6 and 10^7 spores mL^{-1} was safer to predatory mite *P. persimilis* [54]. Number of parasitized larvae of *Eretmocerus sp. nr. furuhashii* survival decreased with increasing concentrations of *L. muscarium* and only 29% emergence of pupae was observed at a conidial concentration of 1 × 10^8 conidia mL^{-1}. Similarly, 67% emergence of adult *E. sp. nr. Furuhashii* was observed [55]. Parasitoid (*Diaeretiella rapae*) emergence was affected by application of *L. longisporum* before or after parasitism and longevity decreased in female F1 populations [56]. In the laboratory conditions, application of *L. muscarium* (1 x 10^8 conidia/ml) against *A. colemani* had not affected longevity and fertility of the female *A. colemani*. The combination of *Aphidius colemani* with *L. muscarium* reduced the aphid infestation in the semi field conditions as compared to *A. colemani* alone [30].

The *Lecanicillium* spp. is not harmful to humans during handling in the laboratory and field for the control of pests.

11. Compatibility of *Lecanicillium* spp. with agro chemicals

Chemical pesticides may have antagonistic or synergistic effect on the potentiality of *Lecanicillium* spp. and may disrupt natural epizootic. Under such epizootic condition, it is expected to enhance effectiveness through joint action of pathogen and compatible insecticides, which would reduce not only the cost of protection but also reduce the contamination of the environment. The literature on compatibility of *Lecanicillium* spp. with agrochemicals is lacking.

Among different insecticides studied for their effect on *L. lecanii* under in-vitro, malathion was significantly detrimental (69.18% inhibition) than all other insecticides except quinalphos (66.76%). Conversely, endosulfon and chlorpyriphos were significantly safer (37.31 to 44.37%), followed by oxydemeton methyl and dimethoate (45.33 to 48.27% inhibition) [4]. Similarly, endosulfan completely inhibited the germination of conidia and hyphal growth. In contrast, diafenthiuron, thiamethoxam, imidacloprid, thiodicarb, primicarb, omethoate, acetamiprid, and pymetrozine were compatible with *L. lecanii in planta* [57]. Imidacloprid and cyromazine were compatible with *L. lecanii* in terms of vegetative growth, sporulation, conidial viability and pathogenicity against *T. urticae.* At the recommended concentration, the fungicides carbendazim, chlorothalonil, propiconazole, mancozeb and wettable sulphur completely inhibited the germination of candida (100%) except iprodione and triadimefom allowed 37.38 and 41.62% conidia to germinate respectively [4].

12. Commercial formulations

The commercial formulations based on *Lecaniillium* spp. are available in India and other countries are presented in **Table 5**. Number of manufacturers based on *Lecanicillium* spp. products is more in India however; the production is very low and not available to the farmers/stakeholders/growers on time as compared to synthetics due to dominant in pesticides market and lack of awareness to farmers/growers about biopesticides. In India, the efficacy of *Lecanicillium* spp., based products was less due to high temperature and low humidity as compared to temperate countries, even though in India, these products were used as one of the components in IPM and also used for the management of sucking pests of flowers and vegetables in greenhouse.

Country	Trade Name	Target pest	Country	Source
Lecanicillium **spp.**				
Honduras, El Salvador, Nicaragua, Jamaica	Verzam	Whiteflies, aphids, thrips, mites	Escuela Agrícola Panamericana Honduras	—
Colombia	Vercani WP	Whiteflies	Colombia	www.ica.gov.co
Uruguay	Lecafol	Whiteflies	Lage y Cía. S.A., Uruguay	www.lageycia.com
L. muscarium				
Denmark, Finland, Italy, UK, Netherlands, Italy, Turkey, Switzerland, Japan, France, India	Mycotal	Whiteflies, thrips	Koppert Biological Systems, Netherlands	www.koppert.com

Country	Trade Name	Target pest	Country	Source
V. lecanii				
India	Bio-Catch	Whitefly, Aphids, Thrips, Mealy bugs	M/s T. Stanes & Company, India	www.tstanes.com
	Multiplex Varsha	aphids, thrips, mealy bug, whitefly, scales mites	Multiplex Biotech Pvt. Ltd., Bengaluru, Karnataka, India	www.multiplexgroup.com
	Verti Guard	---Do--	Lokmangal Bio Tech Maharashtra, India	www.lokmangalbiotech.com
	Sun Bio Verti	---Do--	Sonkul Agro Industries Pvt. Ltd. Maharashtra, India	www.bioorganic.co.in
	Vertisterk	Scales, mealy bugs	Vijaya Agro Industries, Maharashtra, India	www.vijayaagro.com
	Green Basivert	Aphids, thrips, whitefly, mealy bug, scales	Greentech Biotech Laboratory, Tamil Nadu, India	www.agrizone.in
	Vertocoz-P	whitefly, mealy bug	Utkarsh Agrochem Pvt. Ltd., Surat, India	www.utkarshagrochem.com

Table 5.
Commercially available products based on Lecanicillium spp. [58, 59].

13. Conclusions

Lecanicillium spp. is promising biocontrol agent and can be used as one of the components of integrated pest management under green house and field conditions against sucking insect pests. *Lecanicillium* is multiplying on commercially available media (potato dextrose agar and broth etc.) till date but it can be mass multiplied at cheaper rate on solid grain media of sorghum and rice; liquid media of sugar cane molasses. It can be used effectively in conjunction with other natural enemies and compatible pesticides.

Acknowledgements

Author is grateful to Director, Institute of Himalayan Bioresource Technology (Council of Scientific and Industrial Research, New Delhi), Ministry of Science and Technology, Government of India, Palampur, Himachal Pradesh, India for encouragement and support.

Conflict of interest

Author declares that no conflict of interest is reported.

Author details

Sajjalavarahalli Gangireddy Eswara Reddy
Entomology Laboratory, CSIR-Institute of Himalayan Bioresource Technology, Palampur-176061, Himachal Pradesh, India

*Address all correspondence to: ereddy2001@yahoo.com

References

[1] Butt TM, Jackson CW, Murugan W. Fungi as Biocontrol Agents, Progress, Problems and Potentials. CBBS Publishing Co., UK; 2001.

[2] Wang L, Huang J, You M, Liu B. Time dose mortality modelling and virulence indices for six strains of *Verticillium lecanii* against sweet potato whitefly, *Bemisia tabaci* (Gennadius). Journal of Applied Entomology. 2004; 128:494-500.

[3] Gindin G, Barash I, Harari N, Raccah B. Effect of endotoxic compounds isolated from *Verticillium lecani* on the sweet potato whitefly, *Bemisia tabaci*. Phytoparasitica. 1994; 22:189-196.

[4] Bharathi HT. Isolation and characterization of entomopathogenic fungi and their effectiveness [thesis]. Dharwad: University of Agricultural Sciences (India); 2005.

[5] Zare R, Gams W. A revision of *Verticillium* section Prostrata. IV. The genera *Lecanicillium* and *Simplicillium*. Nova Hedwigia. 2001; 73:1-50.

[6] Zare R, Gams W, Culham A,. A revision of *Verticillium* sect. Prostrata. I. Phylogenetic studies using ITS sequences. Nova Hedwigia. 2000;71:465-480.

[7] Gams W, Zare R. A revision of *Verticillium* Sect. Prostrata. III. Genetic classification. Nova Hedwigia. 2001; 72:329-337.

[8] Goettel MS, Koike M, Kim JJ, Aiuchi D. Shinya R, Brodeur J. Potential of *Lecanicillium* spp. for management of insects, nematodes and plant diseases. Journal of Invertebrate Pathology. 2008; 98:256-61.

[9] Shinde SV, Patel KG, Purohit MS, Pandya JR, Sabalpara AN. *Lecanicillium lecanii* (Zimm.) Zare and Games, an important biocontrol agent for the management of insect pests – a review. Agri Review. 2010; 31:235-252.

[10] Kanaoka MA, Isogai S, Murakoshi M, Ichione A, Suzuki A, Tamura S. Bassionalide, a new insecticidal cyclodepsipeptide from *Beauveria bassiana* and *Verticillium lecanii*. Agricultural and Biological Chemistry. 1978; 42:629-640.

[11] Suzuki A, Kanaoka MA, Isogai S, Murakoshi M, Ichione A, Tamura S. Bassionalid. A new insecticidal cyclodepsipeptide from *Beauveria bassiana* and *Verticillium lecanii*. Tetrahedron Letters. 1977;25, 2167-2170.

[12] Soman AG, Gloer JB, Angawi RF, Wicklow, DT, Dowd PF. Vertilecanins: New Phenopicolinic Acid Analogues from *Verticillium lecanii*. Journal of Natural Products. 2001;64: 189-192.

[13] Claydon N, Grove JF. Insecticidal secondary metabolic products from the entomogenous fungus *Verticillium lecanii*. Journal of Invertebrate Pathology. 1982; 40, 413-418.

[14] Gindin G, Geschtovt NU, Raccah B, Barash I. Pathogenicity of *Verticillium lecanii* to different developmental stages of the silver leaf whitefly, *Bemisia argentifolii*. Phytoparasitica. 2000; 28:229-239.

[15] Web of Science 2012 and 2020.

[16] Scorsetti AC, Humber RA, Gregorio CD, Lastra CC. New records of entomopathogenic fungi infecting *Bemisia tabaci* and *Trialeurodes vaporariorum*, pests of horticultural crops, in Argentina. Biological Control. 2008;53:787-796.

[17] Vu VH, Hong SI, Kim K. Selection of entomopathogenic fungi for aphid control. Journal of Bioscience and Bioengineering. 2007; 104:498-505.

[18] Hesketh H, Aldersonb PG, Pyec BJ, Pella JK. The development and multiple uses of a standardized bioassay method to select hypocrealean fungi for biological control of aphids. Biological Control. 2008; 46:242-255.

[19] Diaz BM, Oggerin M, Claudia, Lastra CCL, Rubio V, Fereres A. Characterization and virulence of *Lecanicillium lecanii* against different aphid species. BioControl. 2009; 54:825-835.

[20] Ganassi S, Grazioso P, Moretti A, Sabatini MA. Effects of the fungus *Lecanicillium lecanii* on survival and reproduction of the aphid, *Schizaplis graminum*. BioControl. 2010; 55:299-312.

[21] Subramaniam SR, Babu A, Pradeepa N, A new report of the Entomopathogen, *Lecanicillium lecanii* infecting larvae of tea thrips, *Scirtothrips bispinosus* (Bagnall). Journal of Biosciences Research. 2010; 1:146-148.

[22] Chiriví-Salomón JS, Dannies G, RestrepoS, Sanjuan T. *Lecanicillium sabanense* sp. nov. (Cordycipitaceae) a new fungal entomopathogen of coccids. Phytotaxa. 2015; 234: 63-74.

[23] Lu L, Cheng B, Du D, Hu X, Peng A, Pu Z, Zhang X, Huang Z, Chen Guoqing. Morphological, molecular and virulence characterization of three *Lencanicillium* species infecting Asian citrus psyllids in Huangyan citrus groves. Journal of Invertebrate Pathology. 2015; 125:45-55.

[24] Corallo B, Simeto S, Martínez G, Gómez D, Abreo E, Altier N, Lupo S. Entomopathogenic fungi naturally infecting the eucalypt bronze bug, *Thaumastocoris peregrinus* (Heteroptera: Thaumastocoridae), in Uruguay. Journal of Applied Entomology. 2019; DOI: 10.1111/jen.12624.

[25] Kope HH, Alfaro RI, Lavallee R. Effects of temperature and water activity on *Lecanicillium* spp. conidia germination and growth, and mycosis of *Pissodes strobe*. Biological Control. 2008;53:489-500.

[26] Aiuchi D, Baba Y, Inami K, Shinya R, Tani M, Koike M. Variation in growth at different temperatures and production and size of conidia in hybrid strains of *Verticillium lecanii* (*Lecanicillium* spp.) (Deuteromycotina: Hyphomycetes). Applied Entomology and Zoology. 2008;43, 427-436.

[27] Jaronski ST. Ecological factors in the inundative use of fungal entomopathogens. BioControl. 2010; 55:159-185.

[28] Gao L, Liu XZ, Sun MH, Li SD, Wang JL. Use of a novel two-stage cultivation method to determine the effects of environmental factors on the growth and sporulation of several biocontrol fungi. Mycoscience. 2009; 50:317-321.

[29] Rivas F, Nunez P, Jackson T, Altier N. Effect of temperature and water activity on mycelia radial growth, conidial production and germination of Lecanicillium spp. isolates and their virulence against *Trialeurodes vaporariorum* on tomato plants. BioControl. 2014; 59:99-109.

[30] Mohammed AA, Hatcher PE. Combining entomopathogenic fungi and parasitoids to control the green peach aphid *Myzus persicae.* Biological Control. 2017; 110:44-55.

[31] Ekbom BS. Humidity requirements and storage of the entomopathogenic fungus *(Verticillium lecanii)* for use in greenhouses. Annales Entomologicae Fennicae.1981;47:61-62.

[32] Milner RJ, Lutton JJ. Dependence of *Verticillium lecanii* (Fungi: Hyphomycetes) on high humidities for infection and sporulation using *Myzus persicae* (Homoptera: Aphididae) as host. Environmental Entomology. 1986; 15:380-382.

[33] Kim JJ, Goettel M, Gillespie DR. Evaluation of *Lecanicillium longisporum*, (Vertalec) for simultaneous suppression of cotton aphid, *Aphis gossypii*, and cucumber powdery mildew, *Sphaerotheca fuliginea* on potted cucumbers. Biological Control. 2008; 45:404-409.

[34] Ghaffari S, Karimi J, Kamali S, Mehdikhani E. Biocontrol of *Planococcus citri* (Hemiptera: Pseudococcidae) by *Lecanicillium longisporum* and *Lecanicillium lecanii* under laboratory and greenhouse conditions. Journal of Asia-Pacific Entomology. 2017; 20(2):605-612.

[35] Aiuchi D, Moyo D, Ishikura S, Tani M, Kinoshita R, Rakotondrafara AM, Koike M. Virulence of Lecanicillium spp. (Hypocreales: Cordycipitaceae) hybrid strains against various biological stages of the greenhouse whitefly, *Trialeurodes vaporariorum* (Hemiptera: Aleyroidae). Biocontrol Science and Technology. 2020; DOI: 10.1080/09583157.2020.1771281.

[36] Cuthbertson AGS, Walters KFA. Pathogenicity of the entomopathogenic fungus, *Lecanicillium muscarium*, against the sweet potato whitefly, *Bemisia tabaci* under laboratory and glasshouse conditions. Mycopathologia. 2005; 160:315-319.

[37] Yeo H, Pell JK, Alderson PG, Clark SJ, Pye BJ. Laboratory evaluation of temperature effects on the germination and growth of entomopathogenic fungi and on their pathogenicity to two aphid species. Pest Management Science. 2003; 59:156-165.

[38] Kim JJ, Goettel MS. Gillespie DR. Potential of *Lecanicillium* species for dual microbial control of aphids and the cucumber powdery mildew fungus, *Sphaerotheca fuliginea*. Biological Control. 2007; 40:327-332.

[39] Down R, Cuthbertson AGS, Mathers JJ, Walters KFA.Dissemination of the entomopathogenic fungi, *Lecanicillium longisporum* and *L. muscarium*, by the predatory bug, *Orius laevigatus*, to provide concurrent control of *Myzus persicae*, *Frankliniella occidentalis* and *Bemisia tabaci*. Biological Control. 2009; 50:172-178

[40] Saranya S, Ushakumari R, Jacob S, Philip BM, Efficacy of different entomopathogenic fungi against cowpea aphid, *Aphis craccivora* (Koch). Journal of Biopesticides. 2010; 3:138-142.

[41] Cuthbertson AGS, Walters KFA, Northing P, The susceptibility of immature stages of *Bemisia tabaci* to the entomopathogenic fungus *Lecanicillium muscarium* on tomato and verbena foliage. Mycopathologia. 2005;159,23-29.

[42] Kim JJ, Goettel MS, Gillespie DR. Evaluation of *Lecanicillium longisporum*, (Vertalec) for simultaneous suppression of cotton aphid, *Aphis gossypii*, and cucumber powdery mildew, *Sphaerotheca fuliginea* in a greenhouse environment. Crop Protection. 2010;29: 540-544.

[43] Lee WW, Shin TY, Bae SM, Woo SD. Screening and evaluation of entomopathogenic fungi against the green peach aphid, *Myzus persicae*, using multiple tools. Journal of Asia-Pacific Entomology. 2015; 18:607-615.

[44] Mohammed AA, Kadhim JH, Kamaluddin ZNA. Selection of highly virulent entomopathogenic fungal isolates to control the greenhouse aphid species in Iraq. Egyptian Journal of Biological Pest Control. 2018; 28:71-78.

[45] Xie T, Jiang L, Li J, Hong B, Wang X, Jia Y. Effects of *Lecanicillium lecanii* strain JMC-01 on the physiology, biochemistry, and mortality of *Bemisia tabaci* Q-biotype nymphs. 2019; PeerJ. 7: e7690 DOI

[46] Kamp AM, Bidochka MJ. Conidium production by insect pathogenic fungi on commercially available agars. Letters in Applied Microbiology. 2002; 35:74-77.

[47] Sahayaraj K, Namasivayam SKR, Mass production of entomopathogenic

fungi using agricultural products and by products. African Journal of Biotechnology. 2008; 7:1907-1910.

[48] Derakhshan A, Rabindra RJ, Ramanujam B, Rahimi M. Evaluation of different media and methods of cultivation on the production and viability of entomopathogenic fungi, *Verticillium lecanii* (Zimm.) Viegas. Pakistan Journal of Biological Science. 2008; 11:1506-1509.

[49] Vu VH, Hong SI, Kim K. Production of aerial conidia of *Lecanicillium lecanii* 41185 by solid state fermentation for use as a mycoinsecticide. Mycobiology. 2008; 36:183-189.

[50] Kim JJ, Kim KC, Roberts DW. Impact of the entomopathogenic fungus *Verticillium lecanii* on development of an aphid parasitoid, *Aphidius colemani*. Journal of Invertebrate Pathology. 2005; 88:254-256.

[51] Koike M, Kodama T, Kikuchi A, Okabe M, Kuramoti K, Saito Y. Effects of *Verticillium lecanii* (*Lecanicillium* spp.) against two spotted spider mite, *Tetranychus urticae* and its natural enemy *Phytoseiulus persimilis*. In: Abstracts of 38th annual meeting of society of invertebrate pathology, 7-11 August 2005; Anchorage, Alaska, USA.

[52] Smitha MS, Giraddi RS. Safety of pesticidal sprays to natural enemies in chilli (*Capsicum annuum* L.). J. Biological Control. 2006; 20:7-12.

[53] Derakhshan SA, Ramanujam B. Impact of entomopathogenic fungus, *Verticillium lecanii* (Zimmerman) Viegas on natural enemies of cabbage aphid, *Brevicoryne brassicae* (Linnaeus) and other beneficial insects. Proceedings of the National Symposium on Biological Control of Sucking Pests. 26-27 May, 2006; Bangalore, India.

[54] Donka A, Sermann H, Büttner C. Side-effect of the entomopathogenic fungus *Lecanicillium muscarium* on the predatory mite *Phytoseiulus persimilis*. IOBC/WPRS Bulletin 45;267-269. In: Proceedings of the IOBC/WPRS Working Group "Insect Pathogens and Insect Parasitic Nematodes" and COST Action 862 "Bacterial Toxins for Insect Control, 22-25 June 2009; Pamplona, Spain. Ed. by Ehlers RU, Crickmore N, Enkerli J, Glazer I, Lopez-Ferber M, Tkaczuk C.

[55] Ren SX, Ali S, Huang Z, Wu JH, *Lecanicillium muscarium* as microbial insecticide against whitefly and its interaction with other natural enemies. Current Research, Technology and Education Topics in Applied Microbiology and Microbial Biotechnology. Ed. by Vilas AM, 340-348;2010.

[56] de Silva RJ, Martins ICF, de Alencar JDCC, Silva KP, Cividanes FJ, Duarte RT, Agostini LT, Polanczyk RA. The effect of the entomopathogenic fungus *Lecanicillium longisporum* (Petch) Zare & Gams (Hypocreales: Cordycipitacea) on the aphid parasitoid *Diaeretiella rapae* Mclntoch (Hymenoptera: Braconidae, Aphidiinae). Entomologia Generalis. 2017; 36:219-229.

[57] Gurulingappa P, McGee P, Sword GA. In-vitro and in planta compatibility of insecticides and the endophytic entomopathogen, *Lecanicillium lecanii*. Mycopathologia. 2011; 172:161-168.

[58] Todd KJ, Svircev AM, Goettel MS, Woo SG. The use and regulation of microbial pesticides in representative jurisdictions worldwide [Internet]. 2010. IOBC Global, 99. http:// www. IOBC-lobal.org.

[59] Wraight SP, Lopes RB, Faria M. Microbial control of mite and insect pests of greenhouse crops. 2017; 237-252. In: Lawrence A. Lacey, editor. Microbial Control of Insect and Mite Pests From Theory to Practice. Elsevier; 2017. P. 237-252. Academic Press. DOI: 10.1016/C2015-0-00092-2.

Chapter 11

A Review of Floral Color Signals and Their Heliconiid Butterfly Receivers

Gyanpriya Maharaj, Godfrey Bourne and Abdullah Ansari

Abstract

Signals vary in type and function. However, regardless of the signal, effective transmission and receiver detection are needed to exist for communication. This chapter focuses on a review of visual color signals used by plants to attract pollinators. Signal detection work has intensely focused on epigamic signals; therefore, this review adds to the body of knowledge on nonsexual signal communication. In this review, we investigate visual signals as it relates to pollinators. We focus specifically on visual color signals used by Angiosperms flowers, both static and dynamic, and look at their Heliconiid pollinators as these butterflies provide a perfect organism for studies on floral signal use and pollinators' behavior. We noted that many of these butterflies have three specifically distinct rhodopsins used to identify food and oviposition sites and some have more due to selective pressures of conspecific and mate identification as such they have served as the focal organisms of numerous genetic and ecological studies as they use color signaling in all aspects of their lives. This review further shows that although their color preferences related to feeding, ovipositing, and mate selection have been demonstrated in countless studies, there are gaps in invertebrate literature, as research on the relationships among signal use, evolution, dynamic signals, effects of signals changes on decision making and thus behavior have not been carried out to a large extent.

Keywords: butterfly, Heliconiid, color, pollinator, visual signal

1. Introduction

It is recognized that various signals such as color, sound, vibration, scent, among others, play a pivotal role in attracting animals to con- and heterospecifics within their environment [1]. Receiver's choice is based on an evaluation process whereby these signals are detected and subsequently discriminated [2–6]. Darwin (1871) initially discussed biological signals and their detection in his theories on sexual selection. However, the theoretical framework for the signal detection theory (SDT) was initially developed in 1954 by Peterson, Birdsall, Fox, Tanner, Green and Swets, with Green and Swets [7] going on to develop methods for psychophysics, many of which are used today [8]. With the central strategy of SDT being to manipulate the decision criterion through experiments to expose the sensitivity factors that remain unchanged. More recent work on signal detection encompasses fields from biology to diagnostics and psychology, etc. This review focuses on signal detection theory related to color bias in butterflies, where they are more likely to respond to one color than another. Specifically, this review focuses on color bias of Lepidopteran pollinators and their response to plant signals. As, research on butterfly research is lacking in comparison to their hymenopteran counterparts despite their roles as pollinators and their comparable decline due to habitat loss and land use change [9].

Color is one of the most salient and common signals used in nature for communication within and between taxa as is evidenced in the great diversity in physical appearance of both plants and animals in the natural world [10]. Color and the use of visual displays that incorporate color are used for a wide array of communication and as such influence many behaviors including; foraging, mate recognition and selection, recognition of members of their species and various other forms of inter- and intra-specific communication, such as those between predator and prey and pollinator and plants [11]. Angiosperms, in particular, exhibit many colors and these are often used to communicate with their pollinators [12, 13]. These pollinators, in turn, have complex visual systems that allow for the discrimination of various wavelength of light [2].

Although signal use spans such a wide range, the study of signals in organisms have been very narrow, mainly focusing on sexual selection [14–16]. This chapter is a bibliographic review of over 200 journal articles and books 94 of which are cited ranging from 1919 to 2021 and it aims to add to the body of knowledge on biological signals by focusing on floral color signals used by plants to attract their Heliconiid butterfly pollinators. It specifically focuses on the evolution of visual signals and the use of these signals by these pollinators. It also examines floral color and factors that drive its development and the mechanisms used by these Lepidopteran pollinators to detect this signal, thereby adding to the sparse non-hymenopteran, specifically non-bee, literature available in this area of study.

Heliconiinae or the passion-vine butterfly is a subfamily with in the major family of Nymphalidae. It is one of the best-known butterflies and biologically influential butterflies as it relates to the study of taxonomy, evolutionary biology, mimicry, genetics, coevolution between insects and plants, population biology, animal behavior and conservation biology [17, 18]. This review focuses on these butterflies as they provide a useful system for investigating color signals as butterflies within this subfamily have unique visual systems and use color vision for finding flowers for food, mates and intraspecific communication [19, 20]. Members of the genus *Heliconius* exhibit pollen feeding and as such have developed evolutionary relationship with certain plants as they are dependent on pollen for nutrition, egg production, nuptial gifts, cyanogenesis and increased fitness. Although, the genus does not exhibit many unique structures in comparison to other members of the Nymphalidae family they do have long proboscis with many long bristle shape sensilla trichodea and shorter labial-palpi that help in the collection of pollen grains. They are also able to hold and transport pollen for several hours and over long distances and are efficient pollen harvesters [17]. As such, in addition to pollen feeding, many butterflies within this family are pollinators of many angiosperm families, including Verbenaceae, Cucurbitaceae, Rubiaceae and also Orchidaceae [21]. A study by Maharaj and Bourne show that butterflies specifically *Heliconius melpomene, H. sara and Dryas iulia* [22], see **Figure 1**, are among major visitors and pollinators of *Lantana camara* despite the presence of other viable pollinators in the area including carpenter bees and humming birds.

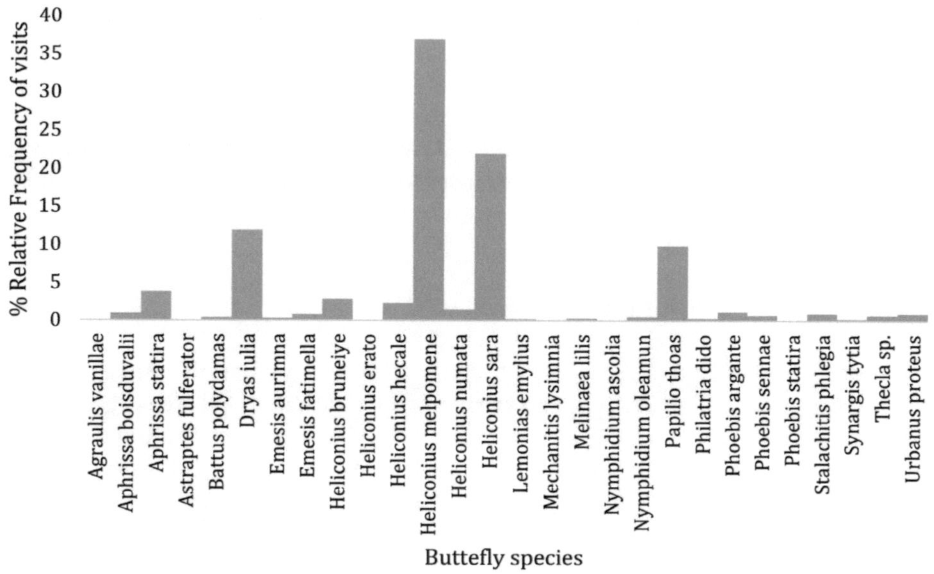

Figure 1.
Frequency of Lepidopteran pollinators observed foraging on L. camara *over a 15-day period. Top three foragers include* Heliconius melpomene, H. sara, Dryas iulia *(modified from Maharaj and Bourne [22]).*

2. Visual signals

Signaling behavior is selected upon only if the signal strength is greater than background noise and can be detected clearly and effectively by receptors [10]. As such, signals, receptors, and behavior are evolutionarily dependent traits and the evolution of one is likely to influence the evolution of the other, as seen in many fishes where visual signals have been noted to evolve in tandem with their visual systems [23]. Often, the environment in which the organism is found, biophysics such as communication between sender and receiver, ability to sensing the environment and foraging choices and, the neurobiological systems of the taxa are all contribution factors driving the evolution of signals, receptors and behavior [10, 24], see **Figure 2**.

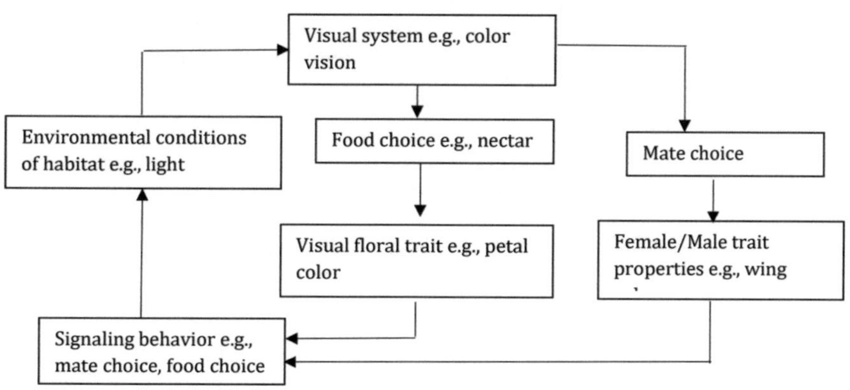

Figure 2.
Process of sensory drive as seen in innate food choices and sexual selection. Arrows indicate evolutionary influences (modified Endler [10]).

Plants signal to a wide range of organisms using many types of visual signals involving both vegetative and reproductive parts [14, 25]. Although we focus this review on flower color and insect attraction, it is recognized that this idea of using floral color signals by plants is not restricted to flowers, as fruits [14] and even leaves [25] exploit insect color preferences. We concentrate on plant-pollinator signals as this provides unique insights into insect-plant communication and a direct way in different aspects of signal theory can be directly tested, such as honesty signals and sensory drive hypothesis.

2.1 Visual signals: Why did they evolve?

The evolution of signals, receptors and signaling behavior stem from the selective pressures exerted by an organism to find food and mates [16], see **Figure 2**. Work by Allen [26], proposed that color vision evolved as a food-finding tool used to locate the edible parts of plants and this led to secondary color preferences such as those for mate attraction and conspecific identification [11, 27]. Ryan and Cummings further link these intrinsic needs by demonstrating that in addition to the cognitive processes of the receiver, such as its preference for a particular trait of its potential mate, there are many organisms in which intraspecific mating preferences can also be influenced by various perceptual biases such as foraging [16]. This type of sensory bias is exploited by male guppies to attract females by using their bias for orange food, water mites that vibrate their legs like prey and male swordtail characins that mimic prey [16, 28–31]. Thus, these senders evolved signals to exploit preexisting biases for food in receivers.

In addition to food, butterflies need visual color signals for mate selection and conspecific identification [32–35]. This is especially seen in Heliconius due to the presences of elaborate Müllerian mimicry rings used in predator avoidance that show a convergence of patterns between close and distantly related species [32–34]. Briscoe and colleagues demonstrate that *Heliconius spp*. mate preference is known to co-evolve with wing color as races are more attracted to their own color patterns [36, 37]. Specifically, *Heliconius spp.* possess positively selected UV opsins that allow detection of distinct yellow colors found on the wings of conspecifics. Additionally, *Heliconius spp*. can use these yellow wing markings to recognize and attract mates; e.g., in *H. pachinus, H. cydno, H. melpomene and H. erato* where females lacking these markings were less attractive to males [36, 37].

Furthermore, it is recognized that organisms also communicate with other completely unrelated taxa. One such relationship is clearly seen in plant-pollinator interactions. Flowers signal presence of rewards through the corolla or other floral parts that are unrewarding [14]. These signals, including flower color, shape, and size, can play an important role in flower detection and choice [38], the is the basis of pollinator syndromes [39].

2.2 Visual signals: How do pollinators interact?

Due to the decoupling of reward and signal in flowers, pollinators must-visit flowers to ascertain rewards offered [14]. As a pollinator approaches a feeding patch it increases its foraging efficiency by making two decisions based on distance. From longer distances a pollinator decides which plants should be approached. And from short distances as they approach the plant, they make the decision on which flower/s should be visited. These decisions are based on the visual attractiveness of plants and flowers, respectively [40]. In many cases, these pollinators display floral consistency by usually visiting one flower per foraging trip even if they routinely collect pollen from multiple sources [41]. This behavior benefits plants by reducing the deposition of heterospecific pollen and increasing conspecific pollen [14] and pollinators by reducing handling times [4]. Another more elaborate form of flower constancy includes traplining which is the collection of food at steady intervals from the same flowers at the same site, thus showing both plant and site faithfulness [42–45] (and citations therein). This behavior has been reported in many taxa, included Heliconius butterflies, although not in great detail, and offers a deeper understanding of floral attraction and pollinators' ability to track rewards offered by flowers displaying honesty signals.

It is posited that animal pollinators' consistency behavior exerts such a strong selective force it is the major driving force behind the diversity in flower color [46]. In one explanatory scenario, it is assumed that each pollinators' behavior is constrained by its limited ability to perceive and distinguish different colors and these constraints vary across taxa. Hence, flower-visiting animals show fixed color preferences, and these preferences differ according to taxa. Therefore, different color signals are aimed at different pollinator groups [47]. An alternate view states that pollinators are relatively unconstrained by their ability to perceive color. Many exhibits true color vision [48], and flower color thus acts as an advertising mechanism to signal visitation induced by the quality of reward offered [46].

Moreover, competition can be a major force in natural selection. As such, exploitative and interference pollinator competition can also contribute to floral divergences in coloration and floral anatomy [49]. As flowers compete for pollinators, pollinators compete for flowers therefore many flowers are visited by several different pollinator species [50]. This in turn leads to resource partitioning by pollinators and assortive mating that in turn leads to floral divergence [51].

3. Floral color

Color signals are an important attractant to pollinators, as flowers, through overt advertising of large brightly color showy petal to a subtle presentation of color combination that acts as guides, communicate with pollinators [52–54]. It is recognized that although color does play an essential part in pollination and this is the focus of this review, its function in plants is not limited to pollinator communication [53, 54].

3.1 Floral color: How is it produced?

Many of the compounds' plants produce are pigmented [55]. Most flower colors are a result of chemical pigments present in the cells of the flower petals. Three major groups of pigments, betalains, carotenoids, and flavonoids, are responsible for the attractive natural display of flower colors [56, 57]. Humans recognize the color of a compound by perceiving reflected or transmitted light of wavelengths between 380 and 730 nm, while insects recognize the light of shorter wavelengths [55].

Betalains, found in the Order plant Caryophyllaceae, are water-soluble nitrogen-containing compounds synthesized from tyrosine by the condensation of betalamic acid, with a derivative of dihydroxyphenylalanine [57]. This reaction results in the formation of the red to violet betacyanins. While the condensation of betalamic acid forms yellow to orange betaxanthins with amino acid or amino acid derivatives [57].

Plant carotenoids, found in a wide array of plants, are 40-carbon isoprenoids with polyene chains that may contain up to 15 conjugated double bonds [58]. They fall into two groups' xanthophylls and carotenes [52] which are the red, orange and yellow lipid-soluble pigments found embedded in chloroplasts and chromoplasts' membranes. These pigments account for the bright colors of fruits and flowers, which often act as attractants to animals [58, 59].

Flavonoids are a large class of secondary plant metabolites of which anthocyanins are the most conspicuous and thus function to attract pollinators when in petals [60]. Flavonoids have a wide range of colors from white, pale yellow to red, purple and blue [56]. Anthocyanins, a less popular group of flavonoids, are responsible for the white, cream to pale yellow coloration of plants that absorb ultraviolet light [52]. They are water-soluble pigments that possess a hydroxylated 2-phenylbenzopyrilium chromophore. There are six types and increases in the number of hydroxyl groups resulting in increases in the visible absorption maximum [56, 61]. Anthocyanins occur in almost all vascular plants' vacuoles and are responsible for the majority of the orange, red, purple, and blue colors of flowers [55, 57].

In addition to pigments, many plants also exhibit morphological characteristics that allow for enhancing the perceived color of the petal. These include, conical or papillate cells found on the petal's adaxial epidermis that increase the amount of light absorbed by the floral pigments [62] found by Kay [63, 64] and later by Glover and Martin and Dyer et al. from experimental evidence from their study of *Antirrhinum majus* that demonstrated that flowers with conical cells received more pollinator attention than those with flat cells [62, 65].

Furthermore, plants also use contrasting floral color traits such as iridescent patches in some orchids, bulls-eye images caused by striations in certain regions of the petal. As exemplified in species such as the *Hibiscus trionum* or darken flower centers as in *Tulipa humilis*. Nectar guides are also seen in many groups which contrast the flower by absorbing light in the UV range thereby increasing the attractiveness of the flower to pollinators by increasing visibility from longer distances and by help animal visitors to orient themselves on the flower prior and post landing [66].

Researchers observed that various floral phenotypes serve to signal or advertise the presence of nutrition rewards [67]. Communication between flowing plants and their pollinators involves a combination of sensory signals that include color, morphology, and odor, which act in concert with each other to become "sensory billboards" [68].

3.2 Floral color: Why did it develop?

One of the most common theories explaining the development and evolution of different floral colors are pollinators as the primary selective agents influencing flower color. Therefore, transitions to different colors represent an adaptation to different suites of pollinators as a selection of one functional group may cause divergence of color while another functional group may maintain that trait [47, 53]. More so, competition for pollinators can account for color divergence as this promotes species level specialization by pollinators, thus decreasing the cost of intraspecific pollen deposition [13].

Initial flower-pollinator observations by Darwin (1862 as cited by Fenster et al. [47]) and many others suggest that different types of pollinators promote selection for diverse floral forms that produce an array of "pollination syndromes," [47]. The primary evidence supporting this contention is the existence of groups of floral traits that occur together associated with attraction and utilization of a specific group of animals as pollinators [47, 53]. As seen in bird-pollinated flowers that are often red or orange with elongated floral tubes, reduced floral limbs, exserted stigmas, and copious dilute nectar as appose to butterfly pollinated flowers which are bright red or orange and have a landing platform and a narrow deep corolla tube, while bee-pollinated flowers, which are typically blue or purple and have short, wide tubes, wide limbs, inserted stigmas, and small amounts of concentrated nectar among many other specialized examples [4, 47, 53].

In addition to the pollinator-shift and the competition models as explanations for floral colors, researchers also recognized the importance of flower pigmentation in other functions aside from visual signaling [54]. For example, enzymes used in anthocyanin synthesis function to make other flavonoid compounds. This in turn affects affect flower color and other ecological and physiological traits such as flower temperature. As such, flower color evolution may be influenced by selection on these pleiotropic effects [53], as flower color mutants not expressing anthocyanins may be less tolerant of stresses such as drought and heat and as such less likely to survive [54]. Other selective pressures such as herbivory also select for flower color, as pigmentation in flowers often correlates with green pigmentation in vegetative tissues, caused by chlorophyll a and b [52], and affect the level of resistance to herbivores [54]. If selection is all together discounted, another view on color divergence is based on the neutrality hypothesis, suggesting that genetic drift is responsible for flower color transitions [53].

3.3 Floral color: How is it used?

Color signals in plants are important to pollinators as they can perceive and distinguish colors and thus show innate and learned color preferences due to reward associations [9]. Flower color can remain constant during the entire anthesis stage, or it can experience color change due to multiple factors such as age, pollinators, or the environment [61, 69–71]. Regardless of if flower color is stable, i.e., remaining one color (as discussed above in pollinator syndromes) or dynamic, i.e., changing during its life span, it functions to communicate with its animal pollinators.

Floral color change (pollination-induced or an age-dependent pattern) has most likely evolved in response to selection by visually orientated pollinators (as was discussed above). It reflects a widespread functional convergence within flowering plants [69]. Von Linne [72] noted that floral color change is a common phenomenon among flowering plants with diverse life histories and growth forms from over 33 orders, 78 families and 250 genera of angiosperms, distributed worldwide, are visited by approximately 15 families of insect and four families of birds [40, 69, 73].

Despite the wide prevalence of flower color change and the well-developed hypotheses offered to explain this trait's adaptive nature, this phenomenon has been experimentally examined in only a few species [40, 73] with results showing varying physiological mechanisms responsible for changes in color such gain or loss of pigments, change in pH, or movement of the floral part such as curling of petals [56, 61, 73, 74]. One of the first theories used to explain red and blue coloration was based on the pH change by Willstatter and Everest [75], where plants would exhibit blue coloration under alkaline conditions and red when acidic [61]. The rivaling theory was by Shibata et al., who proposed the metal complex theory that showed the yellow pigments of plants, flavone, and the flavonal series when reduced with compounds such as sodium amalgamate obtained red anthocyanin solutions [76].

In Angiosperms the location of color changes in fully turgid flowers are dependent on pollinator type [73]. These changes differ in the locations and may affect the entire whorl, several whorls or parts of whorls in combination, or wholly localized to specific areas [73]. For example, plants pollinated by bat or moths generally have color changes in the entire flower; butterflies, bees, and fly pollinated plants usually have localized changes and bird-pollinated flowers can encompass both types of changes [73]. However, regardless of the area affected, it provides crucial information for pollinators that benefit both plant-communicator and animal-receiver with pre-change flowers signaling the provision of rewards and the availability of receptive stigmas. Post-change flowers that are often retained are generally unrewarding and sexually inviable, plants benefit from larger floral displays that attract pollinators over long distances and indicating, at close range, pre-change flowers that are still viable [68, 73, 77]. For example, as seen in Lungworth flowers (*Pulmonaria collina*) which change from red to blue with age [40] or Sweet sage (*Lantana camara*) with one day old yellow flowers offering the heighest wards, while older orange and scarlet flowers offer little or no rewards [69, 78, 79]. The results of a more recent study by Maharaj and Bourne specifically suggested that *L. camara* use two strategies to visually attract pollinator from both short and long distances i.e. 1) honest signaling, as the rewards offered reliably correlated with color stage and 2) billboards communication where multiple colored inflorescences with centrally located scarlet flower buds are surrounded by yellow, orange, red flowers [22]. In addition to color to signal change in reward, plants such as *Quisqualis indica*, with flowers that change color, may be linked with a shift from moth (white flowers) to butterfly (pink/red flowers) pollination [71]. As such, it can be postulated that floral color change is an adaptive trait that benefits both the plant and its insect pollinators by cuing the insects to visit the flowers at the optimal reproductive stage and with the greatest reward [68].

4. Visual systems

Among terrestrial animals, only vertebrates and arthropods have color vision i.e., only these taxa possess the ability to discriminate wavelengths independent of color intensity [48]. One explanation origin of color vision is based the for the selective pressure of an organisms to detect green/yellow and UV wavelengths of light as light reflected from objects are of green/yellow middle energy wavelengths and lacks UV wavelengths. Therefore, if an organism can detect these wavelengths, it can tell the difference between an open space with high UV from a low UV space that can be potential habitats, food or other organisms. This theory is further supported by the presence of UV and green sensitive pigments of primitive arthropods [80].

4.1 Visual systems: What does it comprise?

The compound eyes insects are made up of 8–9 photoreceptor cells surrounded by support and visual pigment cells organized uniquely in optical units called ommatidia [80], see **Figure 3**. Ommatidia are classified as either open, fused, or tiered based on their rhabdoms' structure, which affects the spectral sensitivities of the photoreceptor cells [23]. If open, there is a broader spectral sensitivity, as receptor cells 1–6 each have their rhabdomere that receives its image, if fused, there is narrowing spectral sensitivity as rhabdomeres which have different photopigments act as lateral filters and if tiered, the distal photoreceptor cells filter light from the proximal cells, narrowing the spectral sensitivity [23, 80]. In addition to visual pigments, screening/filtering pigment found surrounding the rhabdom varies in spectral absorption and distribution and affects the eye's spectral sensitivity. However, the interaction between these pigments is not clearly understood [23]. For example, in *Papilio* butterflies, their UV screening pigments superimpose onto their UV or green-sensitive opsins, causing an increase in spectral sensitivity allowing these butterflies to detect six different colors; UV, violet, two kinds of green, and red [80].

Regardless of all the factors that affect the color sensitivity of the eye, for color vision of any kind to exist, opsin genes, which encode visual pigments sensitive to different wavelengths of light, are obligatory [81–84]. Visual pigments are made of two components; a light-sensitive retinal base chromophore (e.g., 11-cis-3- hydroxyretinal) [85] attached by a Schiff- base linkage to an opsin protein [23].

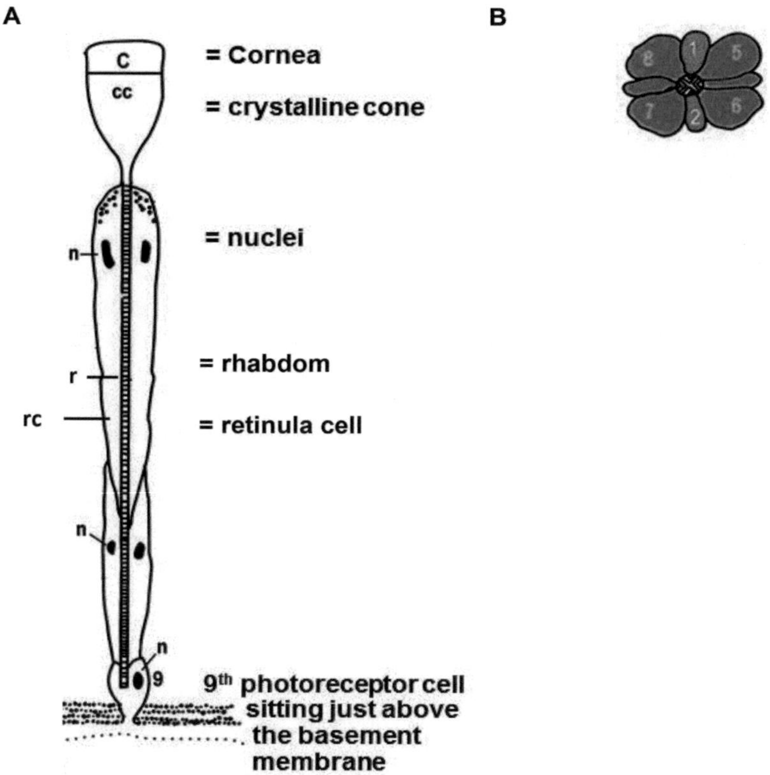

Figure 3.
A. Schematic of an ommatidium. B. Opsin mRNA expression patterns. The cross-sections of three ommatidia are shown. Numbers refer to the photoreceptor cells (R1–R8) (modified from Frentiu et al. [81]).

An opsin belongs to the family of G-protein-couple receptors, and they contain transmembrane domains, which form a binding pocket within which the chromophore is located [23]. Through the chromophore's interaction with critical amino acid residues spectral tuning of the visual pigment to the wavelength of peak absorbance, λ max, is achieved. A diversity of λ max values is achieved through changes in the polarity of amino acids in the chromophore-binding pocket of opsins affect the distribution of electrons in the chromophore π-electron system. However, although the amino acid sequence and the chromophore both affect the maximum absorption λ max, most organisms make a single chromophore, therefore the diversity of the visual pigment absorption spectra primarily depends on the amino acid of the visual pigments [23]. As such, selection for amino acid substitutions at these key sites has led to the spectrally diverse array of visual pigments present in different classes of photoreceptor cells [86]. Thus, these amino acid sites may be under positive selection from selective pressures, such as the organism's light environment, and the need to identify and find food, shelter, oviposition sites (butterflies), mates, and conspecifics [81, 84].

4.2 Visual systems: How did it evolve?

Phylogenetic analyses confirm that opsin genes were duplicated many times before the metazoans' radiations giving rise to several protein subfamilies [81]. In Arthropods, five visual r-opsin families have been identified viz., long-wavelength-sensitive (LW) 1, LW2, middle-wavelength-sensitive (MW) 1, MW2, and short-wavelength-sensitive (SW) [84], with most butterflies possessing three, as in most insects. Peak sensitivities of these opsins include the ultraviolet (UV, 300–400 nm), blue (B, 400–500 nm), and long-wavelength (L, 500-600 nm) part of the light spectrum [87, 88]. Although some butterflies also have a red-sensitive receptor that is also seen in Odonata and Hymenoptera [87].

In bees, moths, and most butterflies, each ommatidium has six or seven receptors expressing long wavelength opsins and two receptors that express two blue and short-wavelength opsins or just one of each [89]. The spectrum visible to butterflies (ultraviolet through the red) is one of the broadest in the animal kingdom [27], making them ideal study specimens in color vision studies. Most butterflies possess the three major spectral classes encoded by ancient duplications, which produced distinct UVRh, BRh, and LWRh opsin genes [27, 84]. Although all butterflies share this similarity, butterfly eyes are incredibly diverse in terms of their spectral organization [48, 90], as some have kept this ancestral arrangement while many other butterflies have many more [11, 91]. For example, swallowtail butterflies *Papilio spp.* have at least three L opsins expressed in the compound eye owing to repeated gene duplication events [92], whereas in the family Pieridae, B opsins are duplicated [93]. Overall, it has been found that representative species of each butterfly family have different numbers of opsins due to lineage specific duplication events of the three basic opsins classes [94]. Butterflies also show diversity in terms of their photopigments' spectral sensitivities and their intraocular filters [11].

Butterflies of the genus Heliconius (Nymphalidae) are considered examples of adaptive radiation due to the spectacular diversity of mimetic wing color patterns that evolved in species and races throughout Mexico and Central and South America [95]. They also have unique visual systems because, besides the pressures of finding food, they must also recognize mates from the multitudinous arrays of mimics [12, 13]. As such, they exhibit remarkable radiation of photoreceptor sensitivities [11]. These butterflies have eyes that contain three or more spectrally distinct rhodopsins, one/two ultraviolet, one blue, and one long-wavelength, as seen in **Figure 3**. Examples are seen in *Dryas iulia*, that have three rhodopsins with λ max = 385, 470, and 555 nm, *Heliconius erato* has eyes that contain four rhodopsins, UVRh1 (UV Rhodopsin 1), UVRh2 (UV Rhodopsin 2), BRh (Blue Rhodopsin), and LWRh (Long wavelength Rhodopsin, with λ max = 355, 398, 470, and 555 nm [90, 94]. This diversity of the eye design reflects the diversity of its evolution and of the lifestyles of the different species as some Lepidoptera use color vision for either feeding, motion vision, oviposition and phototaxis [84, 96]. More specifically, we see a clear link between the evolution of opsins and behavioral preferences e.g., the gene duplication events such as that of the UVRh into UVRh1 and UVRh2 opsin genes have occurred at the same time that UV–yellow pigments of the wings appeared [90] suggesting that the duplicate UV opsin genes has evolved for species recognition and by extension mate selection, in Heliconiid group [90, 94].

5. Conclusion

Generally, photoreceptor sensitivities are adapted for universal vision and do not focus on specific communication signals [11]. However, this is not the case for Heliconiid butterflies that possess a wide diversity of photoreceptors, owing to its multitudinous uses, such as recognition of green leaves for oviposition, yellow, blue, among other color flowers for feeding [84, 97–99], yellow for mate recognition [27, 90, 94] among others.

Bodies of work showing clear-cut evidence for the co-evolutionary relationship between butterfly receptors and mating signals have been substantial. It is also shown that butterflies exhibit innate color preferences associated with feeding [100], and the color of flowers plays a vital role in attracting pollinators [12]. Additionally, Angiosperms employ various strategies to encourage pollinators to approach; color and changing color appear to be particularly important for flower recognition [3, 15]. In particular, the flowers of Angiosperms exhibit tremendous diversity in color that ranges across the UV and visible spectrum [13]. These flowers also differ from pale to nearly black in intensity with closely related sister species or populations of the same species differing in the intensity, hue, or patterning of the corolla [13, 53] caused by numerous evolutionary transitions attributed to pollinator-mediated selection [13, 53].

This review highlights gaps in literature in terms of interrelated research that examine relationships and correlations among communication signals used among and between taxa for conspecific identification, mate selection and plant-pollinator communication, especially in light of Ryan and Cumming's [16] recent review linking the color biases for food and sex in other taxa and van der Kooi et al. [84] demonstrating the clear link with the insect behavior and color vision. It also highlights the need for future research in the field of non-hymenopteran plant-pollinator visual communication and the role changes in color play in conveying messages and affecting decision and subsequently behaviors. This research will facilitate an increase in knowledge in the area of signal theory that has, historically, been biased towards epigamic signals.

Acknowledgements

We wish to thank Dr. Aimee Dunlap, Dr. Yuefeng Wu, Dr. Nathan Muchhala and Dr. Jessica Ware for their guidance during writing.

Notes/thanks/other declarations

This review was constructed from the unpublished Chapter 1 of Gyanpriya Maharaj's PhD dissertation, "Color-mediated foraging by pollinators: A comparative study of two passionflower butterflies at *Lantana camara*" presented to the University of Missouri-St. Louis (UMSL) 12-12-2016.

Author details

Gyanpriya Maharaj[1,2*], Godfrey Bourne[3] and Abdullah Ansari[2]

1 Centre for the Study of Biological Diversity, Faculty of Natural Sciences, University of Guyana, Georgetown, Guyana

2 Department of Biology, Faculty of Natural Sciences, University of Guyana, Georgetown, Guyana

3 CEIBA Biological Center, Madewini, Guyana

*Address all correspondence to: gyanpriya.maharaj@uog.edu.gy

References

[1] Shettleworth SJ. Cognition, evolution, and behavior. Oxford: Oxford University Press; 1990.

[2] Heinrich B. Energetics of Pollination. Annu Rev Ecol Syst. 1975;6(1) 139-170. DOI: 06.110175.001035

[3] Gumbert A. Color choices by bumble bees (Bombus terrestris): Innate preferences and generalization after learning. Behav Ecol Sociobiol. 2000; 48(1):36-43. DOI: 10.1007/s002650 000213

[4] Andersson S, Dobson HEM. Behavioral foraging responses by the butterfly Heliconius melpomene to Lantana camara floral scent. J Chem Ecol. 2003;29(10):2303-2318. DOI: 10.1023/A:1026226514968

[5] Goulson D, Cruise JL, Sparrow KR, Harris AJ, Park KJ, Tinsley MC, et al. Choosing rewarding flowers; perceptual limitations and innate preferences influence decision making in bumblebees and honeybees. Behav Ecol Sociobiol. 2007;61(10):1523-1529. DOI:10.1007/s00265-007-0384-4

[6] Ings T, Raine N, Chittka L. A population comparison of the strength and persistence of innate colour preference and learning speed in the bumblebee Bombus terrestris. Behav Ecol Sociobiol. 2009; 63, 1207-1218. DOI:10.1007/s00265-009-0731-8

[7] Green, D.M., Swets, J.A. Signal detection theory and psychophysics (Vol. 1). New York: Wiley; 1966

[8] Abdi H. Signal detection theory (SDT). In: Encyclopedia of measurement and statistics. 2007; 886-889. DOI: 10.1016/B978-0-08-044894-7.01364-6

[9] Winfree, R., Bartomeus, I., & Cariveau, D. P. Native pollinators in anthropogenic habitats. Annu Rev of Ecol, Evol, and Syst. 2011, 42: 1-22. DOI: 10.1146/annurev-ecolsys-102710-145042

[10] Endler J. Signals, Signal Conditions, and the Direction of Evolution. Am Nat. 1992;139: S125-S153. DOI: 10.1086/285308

[11] Osorio D, Vorobyev M. A review of the evolution of animal colour vision and visual communication signals. Vision Res. 2008;48:2042-2051. DOI: 10.1016/j.visres.2008.06.018

[12] Quattrocchio F, Wing J, Woude K Van Der, Souer E, Vetten N De, Mol J, et al. Molecular Analysis of the anthocyanin2 Gene of Petunia and Its Role in the Evolution of Flower Color. Plant Cell. 1999;11 (8):1433-1444. DOI: 10.1105/tpc.11.8.1433

[13] Muchhala N, Johnsen S, Smith SD. Competition for Hummingbird Pollination Shapes Flower Color Variation in Andean Solanaceae. Evolution. 2014 69(8):2275-2286. DOI: 10.1111/evo.12441

[14] Schaefer HM, Schaefer V, Levey DJ. How plant-animal interactions signal new insights in communication. Trends Ecol Evol. 2004;19(11):577-584. DOI: 10.1016/j.tree.2004.08.003

[15] Pohl NB, Van Wyk J, Campbell DR. Butterflies show flower colour preferences but not constancy in foraging at four plant species. Ecol Entomol. 2011;36(3):290-300. DOI:10.1111/j.1365-2311.2011.01271.x

[16] Ryan MJ, Cummings ME. Perceptual Biases and Mate Choice. Annu Rev Ecol Evol Syst. 2013;44(1):437-459. DOI: 10.1146/annurev-ecolsys-110512-135901

[17] Krenn, H. W., & Penz, C. M. (1998). Mouthparts of Heliconius butterflies (Lepidoptera: Nymphalidae): a search for anatomical adaptations to pollen-feeding behavior. International Journal

of Insect Morphology and Embryology, 1998, 27(4), 301-309. DOI: 10.1016/S0020-7322(98)00022-1

[18] Hsul R, Briscoe AD, Chang BSW, Pierce NE. Molecular evolution of a long wavelength-sensitive opsin in mimetic Heliconius butterflies (Lepidoptera : Nymphalidae). Biol Journal of the Linn Soc 2001; 72(3): 435-449. DOI: 10.1111/j.1095-8312.2001.tb01329.x

[19] Bybee SM, Yuan F, Ramstetter MD, Llorente-Bousquets J, Reed RD, Osorio D, et al. Butterflies Allow a Color Signal to Serve both Mimicry and Intraspecific Communication. Am Nat. 2012;179(1):38-51. DOI: 10.1086/663192

[20] Finkbeiner SD, Fishman DA, Osorio D, Briscoe AD. Ultraviolet and yellow reflectance but not fluorescence is important for visual discrimination of conspecifics by Heliconius erato. J Exp Biol 2017, 220 (7): 1267-1276. DOI: 10.1242/jeb.153593

[21] Corrêa CA, Irgang BE, Moreira GRP. Flower structure of angiosperms used by Heliconius erato phyllis (Lepidoptera, Nymphalidae) in Rio Grande do Sul State, Brazil. Iheringia - Ser Zool. 2001,90:71-84. DOI: 10.1590/S0073-47212001000100008

[22] Maharaj G, Bourne GR. Honest signalling and the billboard effect: how Heliconiid pollinators respond to the trichromatic colour changing Lantana camara L. (Verbenaceae). J Pollinat Ecol. 2017;20(5): 40-50. DOI: 10.26786/1920-7603%282017%29

[23] Briscoe AD, Chittka L. The evolution of color vision in insects. Annu Rev Entomol. 2001;46: 471-510. DOI: 10.1146/annurev.ento.46.1.471

[24] Frentiu FD, Bernard GD, Sison-Mangus MP, Van Zandt Brower A, Briscoe AD, Brower AVZ, et al. Gene duplication is an evolutionary mechanism for expanding spectral diversity in the long-wavelength photopigments of butterflies. Mol Biol Evol. 2007;24(9):2016-2028. DOI: 10.1093/molbev/msm132

[25] Hamilton WD, Brown SP. Autumn tree colours as a handicap signal. Proc R Soc B Biol Sci. 2001;268(1475):1489-1493. DOI: 10.1098/rspb.2001.1672

[26] Allen, G. The colour sense: Its Origin and Development. An essay in comparative psychology. Vol. 10. Boston: Houghton, Osgood & Company. 1879.

[27] Bybee SM, Yuan F, Ramstetter MD, Llorente-Bousquets J, Reed RD, Osorio D, et al. UV Photoreceptors and UV-Yellow Wing Pigments in Heliconius Butterflies Allow a Color Signal to Serve both Mimicry and Intraspecific Communication. Am Nat. 2012.179(1): 38-51. DOI:10.1086/663192

[28] Rodd F, Hughes K. A possible non-sexual origin of mate preference: are male guppies mimicking fruit? Proc Biol Sci. 2002; 269(1490): 475-481. DOI: 10.1098/rspb.2001.1891

[29] Kokko H, Brooks R. The evolution of mate choice and mating biases. Proc Biol Sci. 2003 Mar 22; 270(1515): 653-664. DOI: 10.1098/rspb.2002.2235

[30] Smith C, Barber I, Wootton R. A receiver bias in the origin of three-spined stickleback mate choice. Proc Biol Sci. 2004 May 7; 271(1542): 949-955. DOI: 10.1098/rspb.2004.2690

[31] Bourne G, Watson L. Receiver-bias implicated in the nonsexual origin of female mate choice in the pentamorphic fish Poecilia parae Eigenmann, 1894. Aquarium, Conserv. 2009 2(3): 299-317. https://www.researchgate.net/publication/41506215

[32] Mallet J, Barton N, Gerardo Lamas M, Jose Santisteban C, Manuel

Muedas M, Eeley H. Estimates of selection and gene flow from measures of cline width and linkage dis-equilibrium in heliconius hybrid zones. Genetics. 1990;124(4):921-36. PMID: 2323556; PMCID: PMC1203983.

[33] Jiggins CD, Estrada C, Rodrigues A. Mimicry and the evolution of premating isolation in Heliconlus melpomene Linnaeus. J Evol Biol. 2004;17(3):680-691. DOI: 10.1111/j.1420-9101.2004.00675.x

[34] Klein AL, de Araújo AM. Courtship behavior of Heliconius erato phyllis (Lepidoptera, Nymphalidae) towards virgin and mated females: Conflict between attraction and repulsion signals? J Ethol. 2010;28(3): 409-420. DOI: 10.1007/s10164-010-0209-1

[35] Arikawa K, Wakakuwa M, Qiu X, Kurasawa M, Stavenga DG. Sexual dimorphism of short-wavelength photoreceptors in the small white butterfly, Pieris rapae crucivora. J Neurosci. 2005;25(25): 5935-5942. DOI: 10.1523/JNEUROSCI.1364-05.2005.

[36] Jiggins CD, Naisbit RE, Coe RL, Mallet J. Reproductive isolation caused by colour pattern mimicry. Nature. 2001; 411:302-305. DOI: 10.1038/35077075

[37] Briscoe AD. Reconstructing the ancestral butterfly eye: focus on the opsins. J Exp Biol. 2008; 211:1805-1813. DOI: 10.1242/jeb.01304529.

[38] Waser NM, Price MV. Pollinator Choice and Stabilizing Selection for Flower Color in Delphinium nelsonii. Evolution 2013;35(2):376-390. DOI: 10.2307/2407846

[39] Ollerton J, Alarcón R, Waser NM, Price M V., Watts S, Cranmer L, et al. A global test of the pollination syndrome hypothesis. Ann Bot. 2009; 103:1471-1480. DOI: 10.1093/aob/mcp031

[40] Oberrath R, Böhning-Gaese K. Floral color change and the attraction of insect pollinators in lungwort (Pulmonaria collina). Oecologia. 1999;121(3):383-391. DOI: 10.1007/s004420050943

[41] Cranston PJG and PS. The insects: An Outline of Entomology. 5th ed. Hoboken: Wiley-Blackwell; 2009

[42] Williams NM, Thomson JD. Trapline foraging by bumble bees: III. Temporal patterns of visitation and foraging success at single plants. Behav Ecol. 1998;9(6):612-621. DOI: 10.1093/beheco/9.6.612

[43] Ohashi K, Leslie A, Thomson JD. Trapline foraging by bumble bees: V. Effects of experience and priority on competitive performance. Behav Ecol. 2008;19:936-948. DOI: 10.1093/beheco/arn048

[44] Ohashi K, Thomson JD. Trapline foraging by pollinators: its ontogeny, economics and possible consequences for plants. Ann Bot. 2009; 103:1365-1378. DOI: 10.1093/aob/mcp088

[45] Lihoreau M, Chittka L, Raine NE. Travel optimization by foraging bumblebees through readjustments of traplines after discovery of new feeding locations. Am Nat. 2010;176(6):744-757. DOI: 10.1086/657042

[46] Meléndez-Ackerman E, Campbell DR, Waser NM. Hummingbird behavior and mechanisms of selection on flower color in Ipomopsis. Ecology. 1997;78(8). DOI: 10.1086/657042

[47] Fenster CB, Armbruster WS, Wilson P, Dudash MR, Thomson JD. Pollination Syndromes and Floral Specialization. Annu Rev Ecol Evol Syst. 2004;35(1):375-403. DOI: 10.1146/annurev.ecolsys.34.011802.132347

[48] Sison-Mangus MP, Bernard GD, Lampel J, Briscoe AD. Beauty in the eye

of the beholder: the two blue opsins of lycaenid butterflies and the opsin gene-driven evolution of sexually dimorphic eyes. J Exp Biol. 2006 209:3079-3090. DOI: 10.1242/jeb.02360

[49] Temeles EJ, Newman JT, Newman JH, Cho SY, Mazzotta AR, Kress WJ. Pollinator competition as a driver of floral divergence: An experimental test. PLoS One. 2016. 11(1):e0146431. DOI: 10.1371/journal.pone.0146431

[50] Rodríguez-Gironés MA, Santamaría L. Why are so many bird flowers red? PLoS Biol. 2004;2(10): e350. DOI: 10.1371/journal.pbio.0020350

[51] Rodríguez-Gironés MA, Santamaría L. How Foraging Behaviour and Resource Partitioning Can Drive the Evolution of Flowers and the Structure of Pollination Networks. Vol. 3, The Open Ecology Journal. 2010. 3: 1-11 DOI: 10.2174/1874213001003040001

[52] Kevan PG. Floral colors in the high arctic with reference to insect–flower relations and pollination. Can J Bot. 1972;50(11):2289-2316. DOI: 10.1139/b72-298

[53] Rausher MD. Evolutionary Transitions in Floral Color. Int J Plant Sci. 2008;169(1):7-21. DOI: 10.1086/523358

[54] Campbell DR, Bischoff M, Lord JM, Robertson W. Where have all the blue flowers gone: pollinator responses and selection on flower colour in New Zealand Wahlenbergia albomarginata. J Evol Biol. 2012 Feb;25(2):352-364. DOI: 10.1111/j.1420-9101.2011.02430.x

[55] Tanaka Y, Sasaki N, Ohmiya A. Biosynthesis of plant pigments: anthocyanins, betalains and carotenoids. Plant J. 2008;54(4):733-749. DOI: 10.1111/j.1365-313X.2008.03447.x

[56] Tanaka Y, Tsuda S, Kusumi T. Metabolic Engineering to Modify Flower Color. Plant Cell Physiol. 1998;39(11):1119-1126. DOI: 10.1093/oxfordjournals.pcp.a029312

[57] Grotewold E. The genetics and biochemistry of floral pigments. Annu Rev Plant Biol. 2006; 57:761-780. DOI: 10.1146/annurev.arplant.57.032905.105248

[58] Hirschberg J. Carotenoid biosynthesis in flowering plants. Current Opinion in Plant Biology, 2001;4(3): 210-218. DOI: 10.1016/S1369-5266(00)00163-1

[59] Bartley GE, Scolnik PA. Plant Carotenoids: Pigments for Photoprotection, Visual Attraction, and Human Health. 1995; 7:1027-1038. DOI: 10.1105/tpc.7.7.1027

[60] Holton TA, Cornish EC. Genetics and Biochemistry of Anthocyanin Biosynthesis. 1995;7 (7):1071-1083. DOI: 10.1105/tpc.7.7.1071

[61] Yoshida K, Mori M, Kondo T. Blue flower color development by anthocyanins: from chemical structure to cell physiology. Nat Prod Rep 2009;26(7): 884-915. DOI: 10.1039/b800165k

[62] Glover BJ, Martin C. The role of petal cell shape and pigmentation in pollination success in Antirrhinum majus. Heredity. 1998; 80:778-784. DOI:10.1046/j.1365-2540.1998.00345.x

[63] Kay, Q. O. N. More than the eye can see: the unexpected complexity of petal structure. Plants Today. 1988;109:114.

[64] Kay, Q. O. N., Daoud, H. S. and Stirton, C. H. Pigment distribution, light reflection and cell structure in petals. Bot J Linn Soc,1981; 83: 57-84.

[65] Dyer AG, Whitney HM, Arnold SEJ, Glover BJ, Chittka L. Mutations perturbing petal cell shape and anthocyanin synthesis influence

bumblebee perception of Antirrhinum majus flower colour. Arthropod Plant Interact. 2007;1(1):45-55. DOI: 10.1007/s11829-007-9002-7

[66] Whitney HM, Kolle M, Alvarez-Fernandez R, Steiner U, Glover BJ. Contributions of iridescence to floral patterning. Commun Integr Biol. 2009;2(3):230-232. DOI: 10.4161/cib.2.3.8084

[67] Raguso RA, Willis MA. Synergy between visual and olfactory cues in nectar feeding by wild hawkmoths, Manduca sexta. Anim Behav. 2005;69(2): 407-418. DOI: 10.1016/j.anbehav.2004.04.015

[68] Willmer P, Stanley D a, Steijven K, Matthews IM, Nuttman C V. Bidirectional flower color and shape changes allow a second opportunity for pollination. Curr Biol. 2009;19(11):919-923. DOI: 10.1016/j.cub.2009.03.070

[69] Weiss MR. Floral colour changes as cues for pollinators. Nature. 1991 21;354(6350):227-9. DOI: 10.1038/354227a0

[70] Yoshida K, Miki N, Momonoi K, Kawachi M, Katou K, Okazaki Y, et al. Synchrony between flower opening and petal-color change from red to blue in morning glory, Ipomoea tricolor cv. Heavenly Blue. Proc Jpn Acad Ser B Phys Biol Sci. 2009;85(6):187-197. DOI: 10.2183/pjab.85.187

[71] Yan J, Wang G, Sui Y, Wang M, Zhang L. Pollinator responses to floral colour change, nectar, and scent promote reproductive fitness in Quisqualis indica (Combretaceae). Sci Rep. 2016;6:24408. DOI: 10.1038/srep24408

[72] von Linne C. Philosophia botanica: in qua explicantur fundamenta botanica, 2nd edn. 1763; Vienna.

[73] Weiss MR. Floral color change: A widespread functional convergence. Am J Bot. 1995;82(2):167-185. DOI: 10.1002/j.1537-2197.1995.tb11486.x

[74] Robinson GM. Notes on Variable Colors of Flower Petals. J Am Chem Soc. 1939;61(6): 1606-1607. DOI: 10.1021/ja01875a510

[75] Willstätter, Richard, and Arthur E. Everest. Untersuchungen über die Anthocyane. I. Über den Farbstoff der Kornblume. Justus Liebigs Annalen der Chemie 1913; 401 (2): 189-232.

[76] Shibata K, Shibata Y, Kasiwagi I. Studies on anthocyanins: Color variation in anthocyanins. J Am Chem Soc. 1919;41(2): 208-220: DOI: 10.1021/ja01459a008

[77] Gori D. Floral Color Change in Lupinus argenteus (Fabaceae): Why Should Plants Advertise the Location of Unrewarding Flowers to Pollinators ? Society for the Study of Evolution 1989;43(4):870-881. DOI: 10.1111/j.1558-5646.1989.tb05184.x

[78] Barrows E. Nectar robbing and pollination of Lantana camara (Verbenaceae). Biotropica. 1976;8(2): 132-135. DOI: 10.2307/2989633

[79] Nuttman C V., Semida FM, Zalat S, Willmer PG. Visual cues and foraging choices: Bee visits to floral colour phases in Alkanna orientalis (Boraginaceae). Biol J Linn Soc. 2006;87(3): 427-435. DOI: 10.1111/j.1095-8312.2006.00582.x

[80] Pichaud F, Briscoe A, Desplan C. Evolution of color vision. Curr Opin Neurobiol. 1999;9(5):622-627. DOI: 10.1016/S0959-4388(99)00014-8.

[81] Frentiu FD, Bernard GD, Sison-Mangus MP, Van Zandt Brower A, Briscoe AD, Brower AVZ, et al. Gene duplication is an evolutionary mechanism for expanding spectral diversity in the long-wavelength photopigments of butterflies. Mol Biol

Evol 2007;24(9):2016-2028. DOI: 10.1093/molbev/msm132

[82] Briscoe A. Molecular diversity of visual pigments in the butterfly Papilio glaucus. The Sc. of Nat. 1998; 85(1):33-35. DOI: 10.1007/s001140050448

[83] Koyanagi M, Nagata T, Katoh K, Yamashita S, Tokunaga F. Molecular evolution of arthropod color vision deduced from multiple opsin genes of jumping spiders. J Mol Evol. 2008;66(2):130-137. DOI: 10.1007/s00239-008-9065-9

[84] Van Der Kooi CJ, Stavenga DG, Arikawa K, Belušič G, Kelber A. Evolution of Insect Color Vision: From Spectral Sensitivity to Visual Ecology. Annu Rev Ento. 2021; 66:435-461. DOI: 10.1146/annurev-ento-061720-071644

[85] Smith WC, Goldsmith TH. Phyletic aspects of the distribution of 3-hydroxyretinal in the class insecta. J Mol Evol. 1990; 30(1):72-84. DOI: 10.1007/BF02102454.

[86] Briscoe AD. Functional Diversification of Lepidopteran Opsins Following Gene Duplication. Mol Biol Evol. 2000;18(12);2270-2279. DOI: 10.1093/oxfordjournals.molbev.a003773

[87] Briscoe AD. Reconstructing the ancestral butterfly eye: focus on the opsins. J Exp Biol. 2008; 211:1805-1813. DOI:10.1242/jeb.013045

[88] Bybee S, Yuan F. UV photoreceptors and UV-yellow wing pigments in Heliconius butterflies allow a color signal to serve both mimicry and intraspecific communication. Am Nat. 2012;179(1):38-51. DOI: 10.1086/663192;

[89] Zaccardi G, Kelber A, Sison-Mangus M. Color discrimination in the red range with only one long-wavelength sensitive opsin. J Exp Biol. 2006; 209: 1944-1955. DOI: 10.1242/jeb.02207

[90] Briscoe AD, Bybee SM, Bernard GD, Yuan F, Sison-Mangus MP, Reed RD, et al. Positive selection of a duplicated UV-sensitive visual pigment coincides with wing pigment evolution in Heliconius butterflies. Proc Natl Acad Sci. 2010;107(8):3628-3633. DOI: 10.1073/pnas.0910085107

[91] Lind O, Henze MJ, Kelber A, Osorio D. Coevolution of coloration and colour vision? Philos Trans R Soc B Biol Sci. 2017;372(1724): 20160338. DOI: 10.1098/rstb.2016.0338

[92] Kitamoto J, Ozaki K, Arikawa K. Ultraviolet and violet receptors express identical mrna encoding an ultraviolet-absorbing opsin: identification and histological localization of two mrnas encoding short-wavelength-absorbing opsins in the retina of the butterfly Papilio Xuthus. J Exp. Biol. 2000;203;2894:2887-2894. PMID: 10976026

[93] Awata H, Wakakuwa M. Evolution of color vision in pierid butterflies: blue opsin duplication, ommatidial heterogeneity and eye regionalization in Colias erate. J Comp Physiol A.2009; 195(4):401-408. DOI: 10.1007/s00359-009-0418-7.

[94] Yuan F, Bernard GD, Le J, Briscoe AD. Contrasting modes of evolution of the visual pigments in Heliconius butterflies. Mol Biol Evol. 2010;27(10):2392-2405. DOI: 10.1093/molbev/msq124

[95] Zhang YY, Fischer M, Colot V, Bossdorf O. Epigenetic variation creates potential for evolution of plant phenotypic plasticity. New Phytol. 2013;197(1):314-322. DOI: 10.1111/nph.12010

[96] Awata H, Matsushita A, Wakakuwa M, Arikawa K. Eyes with basic dorsal and specific ventral regions in the glacial Apollo, Parnassius glacialis (Papilionidae). J Exp Biol. 2010;213:4023-4029. DOI: 10.1242/jeb.048678

[97] Weiss MR. Innate colour preferences and flexible colour learning in the pipevine swallowtail. Anim Behav. 1997;53(5):1043-1052. DOI: 10.1006/anbe.1996.0357

[98] Blackiston D, Briscoe AD, Weiss MR. Color vision and learning in the monarch butterfly, *Danaus plexippus* (Nymphalidae). J Exp Biol; 214:509-20. DOI: 10.1242/jeb.048728.

[99] Nuzhnova OK, Vasilevskaya N V. The effect of color preferences on the foraging behavior of the green-veined white butterfly (Pieris napi L.). Contemp Probl Ecol. 2013;6(1):45-50. DOI: 10.1134/S1995425513010113

[100] Hsu R, Briscoe AD, Chang BS, Pierce NE. Molecular evolution of a long wavelength-sensitive opsin in mimetic Heliconius butterflies (Lepidoptera: Nymphalidae). Biol J Linn Soc. 2001;72(3):435-449. DOI: 10.1006/bijl.2000.0511